Strategies That Influence Cost Containment in Animal Research Facilities

Committee on Cost of and Payment for Animal Research
Institute for Laboratory Animal Research
National Research Council

NATIONAL ACADEMY PRESS
Washington, D.C.

NATIONAL ACADEMY PRESS 2101 Constitution Avenue, NW Washington, DC 20218

NOTICE: The project that is the subject of this report was approved by the Governing Board of the National Research Council, whose members are drawn from the councils of the National Academy of Sciences, the National Academy of Engineering, and the Institute of Medicine. The members of the committee responsible for the report were chosen for their special competences and with regard for appropriate balance.

This study was supported by Grant No. N0–0D–4–2139 between the National Academies and the National Institutes of Health of the U.S. Department of Health and Human Services. Any opinions, findings, conclusions, or recommendations expressed in this publication are those of the author(s) and do not necessarily reflect the views of the organizations or agencies that provided support for the project.

International Standard Book Number 0-309-07261-1

Library of Congress Catalog Card Number 00-110818

Additional copies of this report are available from the National Academy Press, 2101 Constitution Avenue, N.W., Lockbox 285, Washington, D.C. 20055; (800) 624-6242 or (202) 334-3313 (in the Washington metropolitan area); Internet: http://www.nap.edu

THE NATIONAL ACADEMIES

National Academy of Sciences
National Academy of Engineering
Institute of Medicine
National Research Council

The **National Academy of Sciences** is a private, nonprofit, self-perpetuating society of distinguished scholars engaged in scientific and engineering research, dedicated to the furtherance of science and technology and to their use for the general welfare. Upon the authority of the charter granted to it by the Congress in 1863, the Academy has a mandate that requires it to advise the federal government on scientific and technical matters. Dr. Bruce M. Alberts is president of the National Academy of Sciences.

The **National Academy of Engineering** was established in 1964, under the charter of the National Academy of Sciences, as a parallel organization of outstanding engineers. It is autonomous in its administration and in the selection of its members, sharing with the National Academy of Sciences the responsibility for advising the federal government. The National Academy of Engineering also sponsors engineering programs aimed at meeting national needs, encourages education and research, and recognizes the superior achievements of engineers. Dr. William A. Wulf is president of the National Academy of Engineering.

The **Institute of Medicine** was established in 1970 by the National Academy of Sciences to secure the services of eminent members of appropriate professions in the examination of policy matters pertaining to the health of the public. The Institute acts under the responsibility given to the National Academy of Sciences by its congressional charter to be an adviser to the federal government and, upon its own initiative, to identify issues of medical care, research, and education. Dr. Kenneth I. Shine is president of the Institute of Medicine.

The **National Research Council** was organized by the National Academy of Sciences in 1916 to associate the broad community of science and technology with the Academy's purposes of furthering knowledge and advising the federal government. Functioning in accordance with general policies determined by the Academy, the Council has become the principal operating agency of both the National Academy of Sciences and the National Academy of Engineering in providing services to the government, the public, and the scientific and engineering communities. The Council is administered jointly by both Academies and the Institute of Medicine. Dr. Bruce M. Alberts and Dr. William A. Wulf are chairman and vice chairman, respectively, of the National Research Council.

JOANNE ZURLO, Center for Alternatives to Animal Testing, Johns Hopkins School of Hygiene and Public Health, Baltimore, Maryland

Staff

Ralph B. Dell, Director
Kathleen A. Beil, Administrative Assistant
Susan S. Vaupel, Editor
Marsha K. Williams, Project Assistant

COMMISSION ON LIFE SCIENCES

Preface

Care and use of animals in research are expensive, prompting efforts to contain or reduce costs. Components of those costs are personnel, regulatory compliance, veterinary medical care, and laboratory animal management, equipment, and procedures. Many efforts have been made to control and reduce personnel costs, the largest contributing factor to cost, through better facility and equipment design, more efficient use of personnel, and automation of many routine operations. However, there has been no comprehensive, recent analysis of the various cost components or examination of the strategies that have been proven or are purported to decrease the cost of animal facility operation.

The National Research Council appointed the Committee on Cost of and Payment for Animal Research (Cost Committee) in January 1998 to examine the current interpretation of governmental policy (Office of Management and Budget Circular A–21) concerning institutional reimbursement for overhead costs of an animal research facility and to describe methods for economically operating an animal research facility. The study was conducted under the auspices of the Institute for Laboratory Animal Research (ILAR) of the Commission on Life Sciences. The committee produced its first report titled *Approaches to Cost Recovery for Animal Research: Implications for Science, Animals, Research Competitiveness, and Regulatory Compliance* in May 1998. The principal conclusion of that report was that animal research facilities are used extensively for the conduct of research and support an environment and animal health profile that are integral to the validity of the experimental animal model. Hence, the facilities and administrative (F&A) costs should be eligible for inclusion in an institution's indirect cost category. The Office of Grants and Acquisition Management of the Department of Health and Human Services ultimately accepted most of this recommendation and extended its applicability to institutions governed by Circulars A–21 and A–122 (see Appendix A). This action also catalyzed an NIH committee's final revisions of the NIH *Cost Accounting and Rate Setting Manual for Laboratory Animal Facilities*. The Cost Committee then considered cost containment methods for animal research facilities and wrote the present report. This report is intended primarily for directors and managers of animal research facilities.

The literature available to the Cost Committee that specifically addresses cost containment methods was relatively sparse. However, two other sources of information were available: The Ohio State University Committee on Institutional Cooperation Study (CIC) of 12 institutions (see Appendix B) and the Yale University 1999 Animal Resources Survey (1999 ARS) of 63 institutions (see Appendix C). The present report is based upon the experience of the committee members, most of whom have been directors of laboratory animal facilities, researchers relying on animal models or professionals overseeing research resources for many years (see biographical sketches, Appendix D), information in the literature, and the two surveys.

This report has been reviewed by persons chosen for their diverse perspectives and technical expertise in accordance with procedures approved by the National Research Council's Report Review Committee. The purposes of the independent review are to provide candid and critical comments that will assist the authors and the National Research Council in making the published report as sound as possible and to ensure that the report meets institutional standards of objectivity, evidence, and responsiveness to the study charge. The contents of the review comments and the manuscript draft remain confidential to protect the integrity of the deliberative process. We thank the following persons for their participation in the review of this report:

Michael Adams, DVM, Professor of Pathology/Comparative Medicine, Wake Forest University School of Medicine, Winston–Salem, NC;
Ronald A. Banks, DVM, Director, Laboratory Animal Resource, School of Medicine, University of Colorado Health Sciences Center, Denver;
B. Taylor Bennett, DVM, PhD, Associate Vice Chancellor for Research, University of Illinois, Chicago;
Linda Cork, DVM, PhD, Chair, Comparative Medicine, Stanford University School of Medicine, CA;
Ron DePinho, MD, Dana-Farber Cancer Institute, Boston, MA;

Robert E. Faith, DVM, PhD, Director, Center for Comparative Medicine,
  Baylor College of Medicine, Houston, TX;
James G. Fox, DVM, Director, Comparative Medicine, Massachusetts
  Institute of Technology, Cambridge;
Warren W. Frost, DVM, MS, Director, Animal Resources Center,
  Montana State University, Bozeman;
Lauretta W. Gerrity, DVM, Director, Animal Resources Program,
  University of Alabama, Birmingham;
Cynthia S. Gillett, DVM, Director, Research Animal Resources,
  University of Minnesota, Minneapolis;
Michael J. Huerkamp, DVM, Assistant Director, Division of Animal
  Resources, Emory University, Atlanta, GA;
Robert O. Jacoby, DVM, PhD, Chairman, Section of Comparative
  Medicine, Yale University School of Medicine, New Haven, CT;
Timothy Kern, PhD, Professor of Medicine and Ophthalmology,
  Director, Center for Diabetes Research, Case Western Reserve
  University, Cleveland, OH;
Dennis F. Kohn, DVM, PhD, Director, Institute of Comparative
  Medicine, Columbia University, New York, NY;
C. Max Lang, DVM, Chair, Department of Comparative Medicine,
  Hershey Medical Center, Pennsylvania State University, Hershey;
Neil S. Lipman, VMD, Director, Research Animal Resource Center,
  Memorial Sloan–Kettering Institute, New York, NY;
Richard J. Rahija, DVM, PhD, Director, Laboratory Animal Resources,
  Duke University Medical Center, Durham, NC;
Irving Weissman, MD, Professor, Department of Pathology, Stanford
  University School of Medicine, CA;
David York, Associate Executive Director for Basic Science, Boyd
  Professor, Pennington Biomedical Research Center, Baton Rouge,
  LA; and,
William P. Yonushonis, DVM, Director, Laboratory Animal Resources,
  Ohio State University, Columbus.

The list shows the diversity and background of the reviewers, again
attesting to the rigor of the process of producing this report. Although the
persons listed have provided many constructive comments and suggestions, responsibility for the final content of this report rests solely with the
authoring committee and the National Research Council.

I am very thankful to the committee members, reviewers, and ILAR
staff. Members of the committee demonstrated their expertise, dedica-
tion, and perseverance and donated their precious time and energy to
focus on this project throughout their tenure on the committee. The

reviewers provided invaluable insights that helped to make the final report more relevant, informative, and robust.

The committee wishes to thank Robert Jacoby of the Section of Comparative Medicine of Yale University School of Medicine, for making available the data from the 1999 ARS, and Rajasekhar Ramakrishnan and Steven Holleran of the Division of Biomathematics and Biostatistics, Department of Pediatrics, College of Physicians and Surgeons, Columbia University, for summarizing and analyzing the data. Ralph Dell was an extraordinary liaison with the groups on the Cost Committee's behalf, playing a pivotal role during our critique and refinement of the survey instrument and the analysis of survey data. The committee deeply appreciated his deft management of the review process and concluding efforts toward publication of the final report. The committee is further indebted to Kathleen Beil and Marsha Williams, of ILAR staff, for their cheerful support of committee functions, manuscript preparation, and producing all the tables (Appendix C) summarizing the 1999 ARS.

Christian E. Newcomer (*Chair*)
Director, Division of Laboratory Animal Medicine
The University of North Carolina

Contents

Executive Summary

The Committee on Cost of and Payment for Animal Research, in the National Research Council's Institute for Laboratory Animal Research (ILAR), was appointed to advise federal funding agencies and grant awardees on three matters:

1. Develop recommendations by which federal auditors and research institutions can establish what cost components of research animal facilities should be charged to institutions' indirect cost pool and what animal research facility cost components should be included in the per diem charges to investigators, and assess the financial and scientific ramifications that these criteria would have among federally funded institutions. The results of this phase of the study were released in an interim report within 6 months of receipt of funding.

2. Determine the cost components of laboratory animal care and use in biomedical research. This will be used to establish a cost baseline that all institutions that use animals in biomedical research, education, and testing can use as a measure of performance efficiency.

3. Assess and recommend methods of cost containment for institutions maintaining animals for biomedical research.

The second task was not done by the committee, because it was discovered that Yale University was well along in planning to conduct a survey of institutions to determine, among other items, cost components of laboratory animal care and use.

1

The Committee on Cost of and Payment for Animal Research used a variety of sources of information in writing this report: the conclusions, but not the underlying data, of a survey conducted by The Ohio State University Office of Research, for the Committee for Institutional Cooperation (CIC study, Appendix B); the 1999 *Animal Resources Survey* (1999 ARS), conducted by the Yale University School of Medicine's Section of Comparative Medicine; published data; and the collective experience of the committee members. The report covers cost of personnel, laboratory animal management, veterinary medical care, equipment and facility design, compliance with regulations, and future directions in research that uses animals.

Of 130 institutions surveyed, 63 responded to the 1999 ARS. To focus on traditional laboratory animal medicine programs, all institutions with an average daily mouse census of 1,000 or more were selected for further analysis. That resulted in 53 institutions that were then grouped by size of mouse holdings: group 1, 1,000-9,999; group 2, 10,000-29,999; and group 3, 30,000 or more.

Personnel represent the largest cost item in the total costs of an animal research facility (ARF), accounting for 50-65% of the total costs. Of the institutions responding to the 1999 ARS 54 had a veterinarian as a director of the animal care program. If institutions with an average daily mouse census of over 1,000 were focused on, there was no difference in mean director full-time equivalents (FTEs) by group size. Furthermore, the institutions in each of the three groups had an average of nearly 1 FTE associate or assistant director and roughly 0.9 FTE business manager. That indicates that directorship overhead was nearly the same regardless of size of institution. Thus, directorship costs per mouse are higher in smaller institutions. Total managerial staff ranged from a mean of 4.0 in group 1 to 5.4 in group 3, again resulting in higher costs per mouse in the smaller group. Total clerical FTEs doubled from group 1 to group 3, and total technical staff rose from 15 to 42 FTEs. In summary, smaller institutions have higher proportional personnel costs, reaffirming the old adage of economy of scale.

As a case study, the use of team management (or "total quality management") at the University of Michigan is described. Animal care has been strengthened and streamlined as a result of having managers, team leaders, and animal care staff work together collaboratively. A more customer-oriented focus has emerged from this process, improving the ability of the animal care program to meet the needs of researchers. Two years after implementation of the team concept, the University of Michigan was able to reduce per diem rates for rodents by 50% and customer complaints dropped to less than half their previous level. Team management improved working conditions, an important factor in staff retention according to the 1999 ARS, although salary and opportunity for advancement were more important retention factors.

Containing costs of laboratory animal management depends on high-quality information yielded by carefully kept records and a comprehensive cost-accounting system. Such a system will permit determination of the costs and benefits of various services and identification of cost savings. It is false economy to purchase animals whose health status and genetic background are unknown; their use can lead to poor scientific data that are inaccurate or misleading because of undetected health problems in the animals. Breeding animals inhouse depends on research needs and on a careful comparison of purchase versus breeding costs. The use of core laboratories is a way to centralize services and thereby realize economies of scale, and it usually results in higher-quality data because core laboratory staff are experienced in the techniques of the laboratory. Such laboratories might produce transgenic or knockout animals, monoclonal antibodies, behavioral testing, and the like.

Costs of veterinary medical care are largely for personnel. The veterinarian director of an animal care program is usually trained in laboratory animal medicine and frequently is a diplomate of the American College of Laboratory Animal Medicine. The salaries of such specialized veterinarians are higher than those of veterinary support personnel, so institutions should make use of these veterinarians to take full advantage of their professional competences and delegate technical and administrative duties to lower-paid employees. Veterinary residents and certified laboratory animal and veterinary technicians can be used as an effective extension of the veterinary medical staff, as noted in the CIC study (Appendix B). Smaller institutions can choose to use part-time veterinary consultants or share positions with other institutions. The mix of species, the presence or absence of a surgery program, and the use of animal models that require intensive veterinary assistance because of experimental complications, invasive procedures, or spontaneous disease are determining factors in the amount of veterinary input required. In general, rodent-only programs require less clinical veterinary support than surgery-intensive programs and programs that use larger species extensively. Well-trained, experienced technicians working under the supervision of a veterinarian can deliver much of the veterinary care required by an institution, thereby lowering costs.

Diagnostic laboratory support is usually contracted for unless the institution is large and can fully support an inhouse laboratory. Health surveillance is expensive, and exact needs depend on several factors, such as species used, source of animals, facility design, and animal housing conditions. Frequency of sampling and method to be used for health surveillance should be based on a risk assessment that incorporates those factors.

The committee considered principles that govern the design of new or renovated animal research facilities, and these principles are presented herein. There are tradeoffs among low maintenance, efficient animal care, investigator convenience, equipment costs, security, and initial cost of construction. Cost estimates are valuable in making choices. Increasing cen-tralization results in increased labor productivity and decreased cost of operation per square foot—a finding that should be considered when renovations or expansions of animal research facilities are contemplated. Decreasing the costs of animal husbandry involves consideration of type of caging (conventional, microisolator, or individually ventilated caging), automatic watering, robot arms for rodent-cage processing, choice of environmental enrichment, bulk purchase of material (depending on space costs), inhouse breeding versus purchase of animals, and medical supplies, including personal protective equipment.

Attention to facility design, equipment, and operating procedures should result in an animal facility that is efficient and easy to manage and maintain. Use of individually ventilated racks could increase intervals between cage changing from 3-4 days to as much as 14 days. Connecting the racks directly to building supply and exhaust can lower maintenance costs by ventilating the cages instead of the whole room. Automatic watering decreases labor costs, but its use can result in undesirable side effects, such as inoperative valves or cage flooding. Using larger water bottles and acidifying or chlorinating the water is an alternative. Careful sizing of animal rooms in the facility permits optimal placement of the racks so that cages can be accessed with a minimum of effort and mobile animal transfer stations can be used. In large facilities, use of robots can permit automation of many parts of the cage-changing process, such as moving cages to the cage-washing room, dumping cages, loading and unloading cages into the cage washer, putting bedding in the cages and filling water bottles, and transporting the clean cages and bottles back to the animal rooms. Experience with the use of robots is limited, and it may be several years before their ability to save costs is determined. Ensuring that the interstitial space (space above the room ceiling) is readily accessible and is laid out so that duct work and machinery are easily maintained reduces costs and exposure of maintenance workers and animals to each other. Walls in rodent rooms might not need to withstand the assault of large animals and can be constructed with material that is less expensive than traditional concrete masonry.

The institutional animal care and use committee (IACUC) is responsible for oversight of an institution's animal care and use program. The cost of that activity is often underestimated because the institution does not account for faculty time spent on IACUC activities. In addition to the costs of faculty time on the IACUC, there are the known costs of administrative staff to support the IACUC functions and the unknown costs of faculty time spent in completing protocols. A National Institutes of Health study of regulatory burden (NIH 1999) cited six major categories of regulatory issues: redundancy of program and facility inspections; different annual reports required by the Office of Laboratory Animal Welfare (OLAW), the US Department of Agriculture (USDA), and the Association for Assessment and Accreditation of Laboratory Animal Care International (AAALAC); USDA requirements that do not allow for professional judgment; significant differences between OLAW and USDA requirements; inconsistent interpretation of regulations and policies by oversight groups; and complexity of regulations governing the import and movement of nonhuman primates. NIH did not estimate the cost of those issues, but addressing them should result in savings of time and money.

Of institutions that replied to the 1999 ARS, 48 reported costs of supporting the IACUC of $0-$301,000. Larger institutions (group 3) spent more on IACUC support, had programs for monitoring use of animals in research in addition to semiannual inspections, and had more faculty and staff serving on IACUCs; but the cost of compliance as a percentage of research dollars received was generally higher for small programs. The proposal to require USDA to regulate use of rats, mice, and birds in research will probably increase the regulatory burden, particularly for smaller institutions.

Many factors will contribute to increased mouse use over the next few years: the genome project and functional genomics, interinstitutional transfer of various mouse lines, conditional and tissue-specific mutations, chemical and viral mutagenesis, creation of therapeutic models, and in vivo gene-transfer experiments. In light of those factors, many institutions are projecting at least a threefold increase over 5 years. Other species—such as rat, rabbit, pig, and nonhuman primate—might become models in gene transfer experiments. In addition, growth in the use of aquatic species—including *Xenopus* frogs, zebrafish, and other fishes—is likely. Such projected increases require construction or renovation of new space, a portion of which must be flexible to accommodate nonrodent species.

Introduction

The Committee on Cost of and Payment for Animal Research, in the National Research Council's Institute for Laboratory Animal Research (ILAR), was appointed to advise federal funding agencies and grant awardees on two matters: (1) Develop recommendations by which federal auditors and research institutions can establish what cost components of research animal facilities should be charged to institutions' indirect cost pool and what animal research facility cost components should be included in the per diem charges to investigators, and assess the financial and scientific ramifications that these criteria would have among federally funded institutions. The results of this phase of the study were to be released in an interim report within 6 months of receipt of funding. (2) Assess and recommend methods of cost containment for institutions maintaining animals for biomedical research.

The first phase of the committee's activities concluded with the publication of the ILAR report *Approaches to Cost Recovery for Animal Research: Implications for Science, Animals, Research Competitiveness, and Regulatory Compliance* (NRC 1998). In that document, the committee recommended that institutions be allowed to recover facilities and administrative (F&A) costs of animal research facilities from the indirect cost pool to be consistent with the allocation of F&A costs for other research space, to ensure high-quality animal-based research, and to ensure humane care of animals consistent with federal regulations.

After publication and public discussion of the committee's report, the Office of Grants and Acquisition Management issued an administrative clarification of Circulars A-21 and A-122 (Action Transmittal OGAM AT 2000-1, dated November 15, 1999, Appendix A) to authorize the allocation of some costs to the F&A cost pool as suggested by the committee. Specifically, those costs were related to procedure rooms, operating and recovery rooms, isolation rooms, quarantine rooms directly related to research protocols, and rooms that house research animals that are not generally removed from the facility for conducting research. Institutions are still required to document, through space surveys, the particular research projects conducted in research space included in the F&A pool. Given those clarifications, an NIH committee completed work on a year 2000 revision of *A Cost Analysis and Rate Setting Manual for Animal Research Facilities* (CARS Manual). The manual was originally produced by NIH in 1974 and revised in 1979. It has been widely used for cost analysis and rate setting in animal research facilities. The 2000 revision of the manual will bring it up to date with federal cost policies and the technical evolution in the animal research facilities.

The ILAR committee's final objective was to analyze the costs entailed in the care and use of animals in biomedical research and to develop useful indicators for institutions to use in scaling their performance efficiency and evaluating their overall support systems for research animals. The committee was also given the charge of assessing and recommending methods of cost containment for institutions that maintain animals for biomedical research. The committee has drawn on a variety of sources to meet its objectives, including published reports in the literature, personal communications with experts in the field, the opinions of the committee's own members, and two survey documents that were available in whole or in part to the committee. The main survey document used by the committee was the 1999 *Animal Resources Survey* (1999 ARS), conducted by the Yale University School of Medicine's Section of Comparative Medicine and analyzed by the Division of Biomathematics and Biostatistics in the Columbia University Department of Pediatrics. Of 130 academic institutions contacted (including the top 100 recipients of NIH funds for 1995), 63 responded to the survey, for a nearly 50% response rate. The total research budget was greater than $50 million for 42 institutions, between $10 and $50 million for 15, and less than $10 million for six. The 1999 ARS questionnaire and a tabular summary of the findings are provided in Appendix C. The survey produced a wealth of descriptive information needed to characterize many variables relevant to contemporary animal care and use programs and practices, but it failed to yield detailed and compelling information about the linkage of costs to the quality of animal care in many areas. Also, a summary of the conclusions, but not the underlying data, of a survey conducted by the Ohio State University, Office of Research, for the Committee for Institutional

Cooperation (CIC study) was available to the ILAR committee for review and consideration. The CIC study included 12 institutions—10 midwestern state institutions and 2 private institutions. Although a small study, it was carefully conducted, with each institution completing a questionnaire and then being visited by an accountant to ensure accurate, high-quality data. This qualitative information is provided in Appendix B to provide readers with an overview of the trends and consequences of various provisions for animal care and use practices in different institutional settings.

Although the approach chosen by the committee has not resulted in the creation of a menu of validated, cost-effective indicators that could predict program excellence or success, it should serve as a useful starting point for institutions involved in planning and conducting cost analyses of their own programs. Institutional philosophy and needs, such as type of barrier housing for rodents and degree of centralization of the animal holding space, have a large impact on costs. Thus, concepts and suggestions made in this report should be used to explore the cost implications of an institution's arrangements for animal care.

It should be noted that although many institutions use the NIH CARS Manual, there remains considerable interinstitutional variation in what is assigned to various cost centers. This variability makes it difficult to compare figures from different institutions and to assess the effectiveness of various cost-saving maneuvers. Furthermore, there is a great reluctance of institutions to share financial data, in that they hold such information to be highly sensitive and confidential. The committee recommends that institutions devote effort to using the newly revised CARS Manual so that the size of various cost centers can be assessed across institutions. A future survey could then collect data on the magnitude of the various cost centers as a function of such variables as species mix, physical plant layout, veterinary services, and personnel mix.

It should also be noted that this report emphasizes containing the costs of using mice in research because they are the most common animal used and, in the experience of the committee, account for a sizable portion of the cost of operating an animal research facility. Furthermore, it is the opinion of the committee that opportunities for cost containment occur most frequently in the care and use of mice. In general, most institutions have witnessed a decline in the use of larger animals (such as nonhuman primates, dogs, cats, pigs, small ruminants, and rabbits) as part of their research portfolio, and costs associated with large animals no longer dominate the total cost of most programs. The cost of care per individual animal of these species has long been known to be high, prompting many institutions to identify the most cost-effective approaches that optimize the care of these species according to the constraints imposed by the institutions' facilities and programs.

Several aspects of a modern ARF are discussed in this report. Personnel costs account for 50-65% of the total costs of an ARF. Hence, a major portion of this report is devoted to reviewing methods of containing personnel costs. Then the cost of complying with regulations is discussed, followed by a consideration of the costs of veterinary medical care. Such issues as veterinary staffing levels and appropriate use of well-trained technicians are considered. Management practices are critical to the efficient operation of an ARF. Administrative aspects of facility operation and animal husbandry practices are both discussed. Impact of facility design on the costs of an ARF is discussed, including some ideas about automation of certain routine tasks. Finally, some ideas about future directions in the use of animals in research are presented and the impact of those research needs on facility capacity and design are discussed.

1

Personnel

Personnel costs are a major component of the cost of operating an animal care and use program, but information generally is lacking on the extent and variation of these costs in different program environments and on useful strategies for cost containment. Adequate staffing is essential to provide high-quality animal care to ensure animal health and well-being, to comply with regulatory guidelines, and to retain public confidence. As emphasized in the *Guide for the Care and Use of Laboratory Animals* (NRC 1996a), the institution should hire sufficient qualified staff to ensure proper care and use of animals in research, teaching, and testing. The factors that influence facility staffing needs include size and type of institution, administrative arrangements for providing animal care and ancillary support activities, physical-plant characteristics, number and species of animals maintained, and the nature of animal research use. Meeting staffing needs is becoming difficult because a high demand for skilled and unskilled labor exists. Furthermore, there is a growing shortage of experienced, trained laboratory animal medicine veterinarians because of increased demand and a decrease in training positions. The 1999 ARS, conducted by the Yale University School of Medicine's Section of Comparative Medicine, does not contain sufficient details to determine a staffing configuration most likely to produce a cost-effective, high-quality animal care and use program in an institution, but it does provide useful information on the general description of contemporary staffing practices and serves as the basis of the committee's comments and recommendations in this regard.

ADMINISTRATIVE PERSONNEL

According to the 1999 ARS, 61 responding institutions have a director and 49 of 63 function with at least a director and a business manager. Many organizations (42 of 63 reporting) also had personnel in assistant- or associate-director positions. In a majority of the 61 organizations with a director, the director was a veterinarian; only seven of 61 institutions indicated that a nonveterinarian held the position of director. That finding reflects the recommendation in the *Guide for the Care and Use of Laboratory Animals* that a veterinarian with training and experience in laboratory animal medicine and science direct a program. With the growth of research animal programs in the last 20 years and the incorporation of technical expertise from research laboratories into centralized research support efforts, the management of personnel, material, physical plant and financial functions has become increasingly complex. That has stimulated the integration of professional managers into the modern research animal organization to allow veterinary professionals to concentrate on scientific collaboration, enhancing research services, advancing the program of veterinary care, institutional interactions, and other dimensions of program direction. Use of full-time or part-time professional business managers is key to the development of sound business practices that could result in significant cost savings.

Veterinarians usually held the positions of assistant or associate director; and in 16 of the 42 organizations reporting in the 1999 ARS, two or more positions were allocated in these job categories. Other types of administrative personnel represented in the survey were, in decreasing order, purchasing agents (30 of 63 institutions), regulatory or compliance personnel (20 of 63 institutions), and informatics specialists (19 of 63 institutions). For each of those job categories, a few institutions had two or more people serving in the position.

In most organizations, according to the 1999 ARS and the CIC Study, personnel costs constitute about 50-65% of the total operational costs of the animal care and use program and are often covered in part by institutional subsidies. This does not reduce an institution's overall cost, but it does reduce the cost base used in the calculation of per diems for cost recovery. Most institutions participating in the 1999 ARS applied subsidies to the support of administrative personnel: 44 of 55 organizations responding indicated that the director's salary was supported at least partially by institutional subsidy. Moreover, 26 organizations provided 100% of the director's salary through institutional funds, and 17 institutions funded an additional one to three professional positions through institutional subsidies. Of the 17, 10 had one additional position, one institution had two, and six institutions had three. Furthermore, 45 of 56 applied subsidies to other professional staff.

Those findings suggest that most institutions appreciate the importance of a sound professional administrative core that provides direction and oversight of their animal care and use program to facilitate animal research activities and to address regulatory compliance. Despite the importance of the senior administrative positions, however, a substantial number of them—43 of 258 (16.7%)—were not filled, according to the 1999 survey. A possible explanation is that institutions are having problems in finding and recruiting qualified personnel or are willing to tolerate vacancies to control costs.

ANIMAL CARE STAFF

The number and quality of animal care personnel are crucial to an institution's ability to maintain the high-quality animal care and use program necessary in today's sophisticated research environment, and institutions appear to make a concerted effort to keep these positions filled. For example, of the 1,413 positions for animal care personnel allocated among the institutions participating in the 1999 ARS, only 71 (5%) were unfilled at the time of the survey.

According to the 1999 ARS, institutions most often use supervisors' assessments to determine appropriate staffing levels for animal care personnel. Time-effort reporting was the second most common method of determination. There are no universally recognized quantitative standards in the field to assist supervisors in determining appropriate staffing levels independently of local facility conditions, species, and types of housing systems. For example, even for a particular caging condition for mice (microbarrier cages with water bottles), the number of cages that technicians were reported to service weekly generally ranged from several hundred to more than 1,200. That suggests that programs wishing to increase cage-change productivity would benefit from exploring such factors as facility design, availability and use of appropriate ancillary equipment, teamwork concepts and division of tasks, and the degree of consolidation of animal populations.

The levels of total managerial and technical staffing dedicated to the animal care functions reported by institutions participating in the 1999 ARS were compared among three groups depending on the size of the mouse population. The 53 institutions that had an average daily census of more than 1,000 mice were divided into three groups depending on the average daily census of mice. Group 1 (23 institutions) had fewer than 10,000 mice each; group 2 (16 institutions) had 10,000 to 30,000 mice; and group 3 (14 institutions) had 30,000 or more mice. There were no statistically significant differences in the average daily census for any other animal species; that strengthens the conclusion that any differences found could be attributed to factors related to differences in mouse census (see Table 10a-d, Appendix C). The total management category consisted of positions described as senior manager, assistant manager, regional supervisor, and training coordinator. The total technical group consisted of positions of animal care technologist, animal care technician, and assistant animal care technician. The means of the full-time equivalents (FTEs) for total managers in the 1999 ARS for groups 1, 2, and 3 were 2.68, 4.58, and 5.95, respectively; and of the FTEs for total technical staff, 15.3, 20.9, and 42.2, respectively (see Figure 1 and Table 8b, Appendix C). Those data from the 1999 ARS show that larger programs realized economies of scale in managerial staffing. The ratio of total technical staff to total animal care management staff was 7.1 in group 3, significantly higher than the 4.6 in group 2 and 5.7 in group 1; 4.6 and 5.7 were not significantly different; large programs reduce costs by having higher technical-to-managerial staff ratios than smaller programs.

PERSONNEL TRAINING

Technician training is important: it produces a competent and efficient workforce that is better able to support an institution's research mission. It can be accomplished through on-the-job training or other inhouse training efforts or through staff participation in a national certification program sponsored by the American Association for Laboratory Animal Science (AALAS). AALAS certification is available on three technical levels: assistant laboratory animal technician (ALAT), laboratory animal technician (LAT), and laboratory animal technologist (LATG). AALAS also confers management certification through its Institute of Laboratory Animal Management.

Of the 63 institutions included in the 1999 ARS, only six did not have any AALAS-certified staff; 488 of 1,573 (31%) people in management, supervisory, and technical positions reported were certified at some level by AALAS. The education required for certification by AALAS enhances the performance of animal care technicians by enabling them to operate with greater technical competence, assume additional job responsibilities, and advance their careers. That statement is supported indirectly by the certification rates calculated by job category in the 1999 ARS. Overall, 172 of 240 (72%) of those in management positions had some level of AALAS certification—65% of senior managers, 83% of assistant managers, 68% of regional supervisors and 100% of training coordinators. Training coordinators had the highest rate of LATG certification (13 of 15, or 87%) followed by senior managers (38 of 72, or 53%). In contrast, only 316 of 1,333

(24%) of those in technical positions were AALAS-certified. In some settings, technical expertise demonstrated by certification has eased the burden of regulatory oversight while bringing greater uniformity to animal care and experimental procedures. For those reasons, institutions should encourage their staff members, through job promotions or other incentives, to participate in the AALAS certification programs.

The increasing sophistication of research animal use and the increasingly complex legislation, guidelines, and policies governing use of animals in research require skilled employees. The use of inhouse resources and mechanisms for training employees might constitute an effective cost-containment strategy by improving the efficiency, effectiveness, and motivation of the work force. According to the 1999 ARS, 89% of the 63 institutions participating in the study had inhouse training programs. In addition to excellent commercially available training materials, a wide array of free materials can be found on the Internet. The latter, found on various university and industry animal care and use program Web pages, can be easily transformed into useful training materials. Cross-training employees is effective in providing diversity to the daily routine and producing a more flexible workforce. Many institutions have noted that well-trained personnel who are cognizant of and engaged in their mission for the institution make a more effective workforce.

TEAM MANAGEMENT: A CASE STUDY

Although widely accepted and practiced in many environments, the application of "total quality management" (or "continuous improvement") concepts to animal care in research institutions is relatively new. On the basis of personal communication with animal care program directors, research institution administrators have recently begun to use team management to organize and manage research animal husbandry; their efficiency has increased, the cost of care has declined, and morale has improved. Because of reports of considerable success, including the experience at the University of Michigan discussed below, this area deserves further study.

At the University of Michigan, the team concept has been used as an animal care management technique for 5 years. There, animal care technicians, animal care managers, veterinary technicians, the veterinary staff, and the administration have, on the basis of customer and staff satisfaction and improved morale, become convinced that it is a superior management method. Although this method might prove to be widely adaptable across diverse recruitment and staffing conditions, it should be noted that attainment of a BS or Associate Degree in Animal Technology was a requirement for employment on the animal care staff at the University of

Michigan. It is interesting to note that 9 of 63 institutions in the 1999 ARS survey offered initial salaries that were higher than the starting salary of $11.25 per hour offered at Michigan.

Some 40 animal caretakers are organized into five husbandry teams. Each team cares for animals in a facility or, in the case of small facilities, in several facilities. One of the teams, the floater team, provides personnel to all teams during member absences or when special projects are conducted. None of these teams include cage-wash personnel, but recently the cage-wash crew has formed a team that includes cage-washers from several buildings. Team leaders meet with the animal care manager and assistant manager once a week for 1-2 hours. Team suggestions and comments are discussed at these meetings, and planning, analysis, and decision-making are based on those suggestions and comments.

Each team has a permanent and a temporary team leader. The temporary team leader is a husbandry technician who has shown promise as a leader and who would like the opportunity to assist in leading the team. Both the permanent and temporary team leaders' duties include training of team members, communicating with investigators, ensuring sufficient supplies, and timekeeping. The temporary-team-leader position rotates every few months, and this provides an opportunity to groom technicians to assume permanent leadership responsibilities. Both team leaders also have daily animal care duties.

Each team meets for a few minutes each morning and has a longer scheduled meeting every 2 weeks. At the morning meetings, adjustments are made in the daily schedule for each team member, especially if some members are absent. At the longer meetings, each team member has an opportunity to place items on the agenda for discussion; the animal care manager, a veterinary technician, a veterinary clinician, and the director or an assistant director usually attends these meetings. The agenda items cover a wide array of topics ranging from animal care standard operating procedures to financial and administrative planning. Team members are encouraged to speak out with no fear of punishment. There is a strong effort to establish consensus regarding new procedures and practices that the team might implement.

The team as a unit is responsible for all aspects of animal care in the facility or facilities assigned to the team. Workload is apportioned to the members of the team through mutual consent of the members. Requests for additional personnel come from the team. Each member has a stake in the successes or failures of the team, and all members participate in problem solving when new challenges or opportunities are placed before the team. As team management concepts have become more accepted, managers, team leaders, and animal care staff have undergone shifts in outlook that have strengthened and streamlined animal care. The managers and team leaders see themselves as leaders and coaches more than as managers and controllers. The animal care technicians see themselves more as partners that are empowered to shape the work. Problem solving has become a unifying experience, and the teams have taken on a more customer-oriented focus. Cooperation and participation have become normal, and more energy is focused on meeting needs of the researchers. Turnover rate among animal care technicians at Michigan is high for two reasons: first, some leave to take a position that uses more of their BS training; second, some are hired by the scientific laboratories to manage animal-using activities. Two years after implementation of the team concept, the University of Michigan was able to reduce per diem rates for rodents by 50%, and customer complaints dropped to less than half their previous level.

Organization of husbandry has been so successful that several other groups in the animal facility have also organized themselves into teams. These groups include the veterinary medical care team, the administration team, and the institutional animal care and use office team.

The university strongly supports team management by providing team-leader training and providing facilitators to assist teams in organizing. The university also provides awards for the best team effort campuswide. The university administration sees the principal goals of team management as respecting people and ideas, managing by fact, and satisfying customers.

SALARIES, BENEFITS AND INCENTIVES

The 1999 ARS explored many aspects of staffing of animal research facilities. Animal care managers and others might find it helpful to compare the survey responses to the situation in their institutions (Table 8f, Appendix C). In the surveyed group, the standard workweek was 39.3 hours (range, 32.5-42 hours). The average entry-level hourly wage for animal care staff was $9.05 (range, $6.02-$14.14). The average annual salary for animal care staff as a whole was $22,268 (range, $15,149-$34,000). Fringe benefits averaged 26.6% of salary (range, 14-39%). A possible explanation for the observed variation is region-to-region variation in labor availability and prevailing salaries. At the 23 institutions where animal care staff were all or mostly unionized (Table 8d, Appendix C), mean direct salary was $23,697; at the 31 institutions where staff were largely or completely nonunionized, annual salary was $21,173, a statistically significant difference ($p<0.05$). In the institutions surveyed, the mean number of vacation days for animal care staff was 15.6/year, plus 11.9 paid sick days, 9.7 paid holidays, 0.9 other recess days, and 1.6 personal days, for a total of nearly 40 days/year.

Recruitment and retention of animal care technicians have become major issues for most institutions. Animal care managers were asked (1999 ARS) to rank a variety of factors that were potentially important in recruitment and retention of personnel as high, moderate, low, or no importance (Table 8g, 8h, Appendix C). For recruitment of animal technical staff, starting salary and earning potential were ranked as highly or moderately important in 68% of the 53 institutions that used mice, while benefits were highly or moderately important in only 25% of institutions. Recruitment of trained, experienced staff members was seen as highly or moderately important by 66% of the 53 institutions. Job responsibility, career opportunities, regional competition, and geographic location were highly or moderately important in recruitment in 53%, 60%, 57%, and 42% of the institutions, respectively.

With respect to retention of animal care technicians, animal care managers rated earning potential as the most important factor (70%) followed by career opportunity (65%), regional competition (62%), working conditions (53%), and benefits (25%) (Table 8i, 8j, Appendix C). Retention of animal care technicians is important because well-trained, experienced animal care technicians are key to a program's ability to deliver efficient and quality service. High turnover ratios are expensive because of high training costs and lack of productivity of newly hired technicians.

OUTSOURCING ANIMAL CARE SERVICES

Outsourcing, the use of leased labor, is used as a strategy in some organizations to attain labor-cost savings and unburden internal administrative, supervisory, and regulatory systems. Only three of the institutions participating in the 1999 ARS reported having experience with outsourcing, so the evaluation of this strategy as an effective cost-containment method is not possible. Use of outsourcing is more widespread among government agencies that have animal care and use activities and in the industrial laboratory animal sector. The benefit of this approach is that it allows an institution to maintain a specialized labor pool with defined job qualifications, higher commitment and productivity, and lower turnover rates than might be achieved through internal administrative-personnel recruitment and development mechanisms (Houghtling 1998). There are anecdotal reports that—through skillful contract negotiation, clear benchmarking, and careful attention to approval of overtime requests—institutions have been able to effect substantial labor-cost savings and assemble an effective and well-qualified workforce by outsourcing. However, published information on this approach in the laboratory animal industry is insufficient to support a recommendation.

SUMMARY

In summary, the major findings and opinions expressed in this chapter are as follows:

- Most institutions maintain and subsidize a critical administrative nucleus of professional veterinary and/or management personnel involved in program oversight. The data from the 1999 ARS did not permit the evaluation of the administrative configurations against program quality performance measures. The vacancy rate for these positions was 16.7%, suggesting the need for enhanced development, recruitment, and retention efforts to ensure sound program leadership.
- Large mouse-based animal care and use programs are able to operate with higher ratios of technical staff to animal care management staff and so to realize an economy of scale in managerial staffing.
- Inhouse training was the predominant mode (89%) used for preparing the workforce among the institutions participating in the 1999 ARS. Certification at some level by the American Association for Laboratory Animal Science was more prevalent among management positions (72%) than among technical positions (24%).
- The application of the team management approach (University of Michigan study) suggests that institutions should be encouraged to apply modern management techniques to enhance investigator (customer) satisfaction, improve employee performance and involvement, and potentially reduce costs. This approach may be more easily implemented by hiring and retaining employees with training and skills in personnel management.

2

Laboratory Animal Management Practices

ADMINISTRATIVE PRACTICES

Records

A good record-keeping system is important for the efficient operation of an animal research facility (ARF). Records that must be kept by an ARF are of three general types, namely, animal records, financial management records, and compliance records. Animal records contain such information as the source of the animal; the animal's species, strain, gender, and any other pertinent characteristics; the date of receipt of the animal; and the date and nature of the animal's final disposition. Animal records must also identify protocols on which the animal is used and diagnostic and medical procedures used on the animal. To reduce the labor requirement and cost of animal record-keeping, a single record may cover homogeneous groups of animals. For example, a group of animals from the same source, of the same strain, received on the same date, housed in the same room, subject to the same diagnostic and medical procedures, and used on the same protocol can be covered by a single record with a notation of the number of animals involved. Basic to animal records is accurate animal identification. Animal facility management and investigators should evaluate and agree on appropriate animal identification methods and see that they are implemented consistently and conscientiously. Inaccurately identified animals can lead to inaccurate data, which can lead to the costly need to repeat experiments.

Financial management records are necessary for cost analysis and the recovery of ARF costs through fees for services. These records include census records on the number of animals per day assigned to an investigator or protocol. They must also include the billing and payment records of investigators or protocols. Cost analysis records include personnel activity reports or other data for allocating salaries and wages to animal categories, space use records by animal category, cage-washing schedules and the number of cages washed by animal category, the quantities and costs of supplies used by animal category, and the cost of animals procured. Additional records might be necessary for accurate cost analysis, and the reader is referred to the *Cost Analysis and Rate Setting Manual for Animal Research Facilities* (CARS Manual) (NIH 2000 or http://www.ncrr. nih.gov/) for such information. Data collected for cost analysis should be examined to see whether they reveal opportunities for cost containment. For example, personnel activity reports could reveal inefficient assignment of personnel, and revision of assignments could lead to cost savings.

Compliance records are those required for compliance with the Animal Welfare Act, the Public Health Service Policy on the Humane Care and Use of Laboratory Animals, and any other applicable laws and regulations. Included in these records are those of protocol reviews and approvals, numbers of animals and species approved and used for a protocol, and reviews of animal care and use programs and facilities. Also pertinent are occupational health, faculty and staff training, and facility security records.

Records are essential but can be a substantial cost item for an animal research facility. The institution must give thought to the type and format of records and the intended uses of the data collected. Data should not be collected and recorded unless the institution foresees a need for the information. Similarly, records should not be retained beyond their useful life. Note that some compliance records must be retained for 3 years after termination of the research project. There is a large amount of interrelationship among the records kept by an animal research facility. For example, the number of animals procured and assigned to a protocol needs to be entered into animal records, financial management records, and compliance records. Because of this interrelatedness, the institution should set up a system of interrelated databases to minimize data entry.

Cost Accounting

Cost accounting is very important for the efficient and cost-effective operation of an ARF. The facility should have a system of cost accounting like that described in the CARS Manual. This manual sets forth a method whereby an ARF can allocate its costs to specific animal categories and service activities. The total costs associated with an animal category or service divided by the number of animal days or service units yields per diem or service unit costs. The manual contains a discussion of how such unit costs can be used to determine fees charged to users. Fees determined by these methods can be explained to any interested investigator. Investigator understanding of the costs involved in the care of their research animals generally leads to a greater acceptance of those fees. Fees based on cost accounting are more readily justified to sponsors of research. The cost analysis and related statistical data also will assist an institution in comparing the costs and benefits of various services and activities and have the potential for identifying how cost savings might be achieved. Cost records can also be used to develop cost consciousness in the entire staff. A sense of pride in being part of an efficient facility is a useful element in controlling costs. The major cost components of animal care are listed in Table 1.

Almost all institutions have a system of charges for services to support their animal research facilities. As noted in the CIC study, nearly all institutions provide supplemental support from institutional funds. Per diem charges for animal care generally include housing, husbandry, cage sanitation, and maintenance of census records; in most institutions, routine medical care is also included. Routine veterinary medical care includes rodent health surveillance (sentinel animals, bedding transfer to sentinels, serology, and necropsy), disease diagnosis, physical examination of nonrodent mammals on arrival, response to medical emergencies, clinical and anatomic pathology support of diagnosis, and pharmacy stocking and maintenance. At the University of Michigan, each of these activities is attributed to an animal species in proportion to use of the

TABLE 1   Relative Components of Animal Care Per Diem (1999 ARS[a])

Component	Fraction of Per Diem Cost, %
Husbandry	51
General and administrative	15
Cage washing and sanitation	12
Maintenance and repair	6
Health care	5
Laboratory services	4
Technical services	2
Transportation	1
Training	1
Receipt/processing	1

[a]Taken from Table 20b, Appendix C.

activity for the purpose of setting the veterinary service fee (VSF) portion of the per diem. Every investigator at the university pays the daily VSF for his or her animals no matter who provides the daily care. For example, in the 1999 ARS, 74% of the institutions included support for routine rodent medical care in their per diems, and the remainder had a special fee (Table 14a, Appendix C). However, 63% of the institutions had a special fee for therapy of protocol-related disease. Institutions frequently provide a range of technical services on a fee-for-service basis: 42% had special fees for rodent euthanasia, 49% for rodent identification, 55% for rodent special diets, 56% for rodent breeding, 76% for rodent restraint, 89% for specimen collection, 88% for compound administration, and 76% for rodent rederivation (Table 12a, b, & c, Appendix C). A mixture of per diem and direct service charges makes good sense in that the user pays for special services.

Animal Procurement

Animals of the appropriate species, genetic makeup, and quality must be procured for research purposes. Purchase of animals with uncertain health and unknown genetic background constitutes false economy in that their use can lead to inaccurate and invalid data or the necessity to repeat experiments. The decision to breed animals inhouse or to obtain them from commercial sources can be made after a careful analysis of all relevant factors. These include the purchase and shipping costs for commercial animals, the cost of inhouse breeding (including space costs), and the reliability of animal supply and quality.

Research Services

For efficient animal research, an institution can provide central core laboratories for a number of services rather than having individual laboratories duplicate services. These can be "free-standing" core laboratories or be provided by a laboratory otherwise heavily engaged in that activity.

An example of one such service is cryopreservation of embryos. It is expensive to maintain breeding colonies of mutant mice or mice whose genome has been genetically manipulated unless there is an immediate need for them. It is often desirable to maintain unique genetic material or protect it against loss; at present, this can be done most economically by cryopreservation of embryos, but methods for the cryopreservation of rodent semen are also under development and might be applicable to some models. In the 1999 ARS, many institutions reported making cryopreservation of embryos or sperm available (Table 16e, Appendix C). In particular, 78% of group 3 institutions (730,000 mouse average daily census) reported making cryopreservation available through the animal resource program or other internal source. In this group, 43% of the institutions asked the investigators to bear the expense.

It also might be desirable to establish specialized core laboratories for other activities, including monoclonal antibody production, production of transgenic or gene-knockout animals, characterization (by organ system or clinical specialty) of the phenotype of induced mutations in mammals, behavioral testing, histopathologic analysis, and experimental surgery (Tables 16a-h, Appendix C). Experimental surgery and, in larger programs, histopathology services are generally provided by the ARF, whereas other core services are generally provided by other internal sources or an external vendor.

Physical Plant

The physical plant of an animal facility must be designed to maintain the proper environment for the animals and to facilitate the investigative use of the animals. A well-designed physical plant with low maintenance costs, providing for efficient animal care and effective use of the animals by investigators, is an important element in controlling costs. Admittedly, there can be tradeoffs among low maintenance, efficient animal care, investigators' convenience, and the initial cost of construction; these factors will vary institution by institution, and careful analysis should be given in each situation.

There is a clear economy of scale in animal research facilities. The CIC study findings (Appendix B) indicated that labor productivity was the prime driver of animal care costs. Labor productivity was better in larger facilities. For example, caretaker productivity doubled when the labor-weighted volume (adjusting for the labor component of care across different species) increased fivefold. When an institution had more centralized facilities, labor productivity increased. For example, institutions with one or two facilities had a labor-productivity index about 1.5 times greater than institutions with 14 or more sites. Analysis of 1999 ARS confirmed and extended findings of the CIC study. There were 44 respondents who provided sufficient information to compute total operating cost of the facility and who listed the number of sites in their facility by size category (<5,000, 5,000-10,000, 10,000-20,000, and >20,000 ft^2, Table 4, Appendix C). Total facility costs were regressed on amount of space (in square feet) in each category. Costs in dollars per square foot dropped from \$93/ft^2 in the second category (5,000-10,000) to \$36/ft^2 in the third and \$28/ft^2 in the fourth (Figure 2). The differences between those values were statistically significant at $p < 0.0001$; the coefficient for the smallest category was not statistically significant. Labor productivity also increased as caretaker hours per room increased. For example, labor productivity doubled when annual caretaker hours per room increased from 100 to 400 (CIC study, Appendix B). Those findings support the recommendation that animal care operations be concentrated, whenever possible, into fewer larger sites. Concentration of animal facilities must be weighed against investigator convenience in having animals readily available.

Security is a major concern for animal research facilities and can constitute a substantial cost item. The 1999 ARS indicated that institutions had 46% of their sites protected by locks and keys, 17% by electronics, and 37% by a combination of electronics and locks and keys. Institutions should give careful attention to the risk of intrusion and the costs and benefits of various security systems. In addition to the economic costs, institutions should recognize that the public relations and psychologic costs of unwanted intrusions into an animal facility or research laboratory can be substantial.

It must be recognized that the physical plant of an animal facility is a "hard use" area. The sanitizing materials, high traffic, heavy rolling equipment, the active nature of animal care and use, and some animal species themselves all exact a toll on the physical facility. That toll and the requirement to maintain reliable heating, ventilation, and air conditioning, electric systems, and sanitation and sterilization equipment dictate the need for constant maintenance. It is frustrating to the animal care staff, inefficient for operations, and a detriment to quality research when aspects of the physical plant underperform or require frequent maintenance. A well-maintained physical plant in which all systems operate reliably contributes to cost-efficient animal care.

Nearly all institutions use 100% outside air with no recirculation. Because the air is conditioned (heated and humidified or cooled), not recirculating the conditioned air is expensive. According to the *Guide,* some recirculation is possible if the recirculated air is appropriately treated to remove microbial and chemical contaminants. Another method of energy recovery is to use heat exchangers to partially heat or cool the incoming outside air.

ANIMAL HUSBANDRY

Mouse Husbandry

The current methods of mouse husbandry were given considerable attention in the 1999 ARS in acknowledgment of the emerging prominence of mouse models in contemporary biomedical research. Nearly all institutions (98%) were housing some mice in microbarrier cages. Only a single small institution had not implemented microbarrier housing.

Most institutions (67%) were using some individually ventilated cages. More large institutions (79%) were using these cages, whereas only 52% of smaller institutions were using this newer labor-saving cage system. Table 11a-11d, Appendix C, contains information on the 53 institutions with a mouse average daily census of more than 1,000.

Automatic watering systems for mice have been controversial both because some mice develop dehydration if unable or untrained to manipulate the valves properly and because cages can be flooded if an automatic valve leaks or is continuously manipulated by the mice. Only 41% of the surveyed institutions had any automatically watered cages (Table 11a, Appendix C). Fewer group 1 (32%) and group 3 (29%) institutions used automatic watering systems than group 2 (67%) institutions. The 1999 ARS did not explore the role of such factors as cost, customer satisfaction, criteria for selecting a particular system, ease of sanitation, efficiency of operation, and intensity of oversight necessary to ensure proper function in the decision to deploy these systems.

Some institutions house mice in microbarrier cages but do not use HEPA-filtered change hoods for transferring mice to clean cages. The percentage of mice changed in HEPA-filtered change hoods averaged 55% in small institutions, 76% in medium institutions, and 61% in large institutions (Table 11a, Appendix C).

Microbarrier cages are changed more frequently than open-top cages because of ammonia accumulation. The average interval between changes in microbarrier cages was 5.4 days in small institutions, 4.6 days in medium institutions, and 5.9 days in large institutions, with a range of 3-7 days (Table 11a, Appendix C). The survey showed that cage-changing was less frequent in individually ventilated cages; however, the mean interval between cage changes was much smaller in practice than commonly advertised for these systems. The mean interval between cage changes in individually ventilated cages averaged 8.2 days in small institutions and 8.9 days in medium and large institutions; the range for all institutions was 3.5-14 days.

A summary of the CIC study findings (Appendix B) indicated that the cost of animal care is lower in rooms that house larger numbers of animals. In the 1999 ARS, institutions were asked about the maximal number of adult mice that were permitted in their standard shoebox cages presumed to provide about 70-75 in.^{2} of floor space. Most institutions (66%) permitted five mice per cage; 29% permitted only four mice per cage (Table 11b, Appendix C).

The average number of mouse-cage racks in a room was 4.1; the range was 2-8 (Table 11b, Appendix C). In these institutions, respondents were asked about the minimal aisle width that they recommended between racks. The average of the responses was 3.1 ft; the range was 0.5-8 feet. In the experience of the members of this committee, few animal care technicians or research technicians are comfortable in performing animal room duties in aisle widths below the mean reported in the ARS; this is also reported in the case study in this report (Chapter 4). The consensus of the committee was that room design, ergonomic considerations, heating, ventilation, and air conditioning capacity to maintain appropriate ambient air conditions should be evaluated by each institution to preserve a high-quality work and research animal environment before pursuing higher room capacities as a strategy for cost containment.

To reduce expenses, facility managers are exploring different methods of sanitizing mouse cages. The 1999 ARS (Table 11c, Appendix C) indicated that 81% of the institutions were autoclaving their microbarrier cages, 49% were autoclaving their individually ventilated cages, and only 8% were autoclaving their open-top conventional cages; 19% only autoclaved cages used for immunodeficient mice; and a few (8%) used hot water without detergent to clean cages before autoclaving them. The type of cage washing and autoclaving used by an institution will depend on the microbiologic status of the mice housed in the facility.

Various methods of bedding disposal were used (Table 11d, Appendix C). Most institutions (75%) disposed of soiled bedding in a landfill, 26% disposed of soiled bedding by incineration, and 21% disposed of some soiled bedding in the sanitary sewer. Nearly all institutions disposed of animal carcasses by incineration; only 8% reported landfill disposal.

Cost Containment

The scope of animal-husbandry activities required to support biomedical research is extremely diverse because of the wide variety of animal species used and the requirements of the varied research being performed. Those factors make it difficult to identify cost-saving measures that will apply universally. Some general observations regarding cost considerations and potential savings are presented here with respect to common areas of animal husbandry, such as cage sanitizing, watering, environmental enrichment, purchasing supplies, and acquiring animals.

Cages and Cage Processing

Transferring animals to clean cages and sanitizing primary enclosures constitute the bulk of physical labor required to support research facilities that have large rodent populations. It is important to schedule these activities carefully so that staff changing cages have clean cages and equipment (water bottles, cage tops, card holders, and so on) as they are needed and staff washing cages can plan activities in the wash room. Several innovations show promise for minimizing costs associated with these husbandry requirements. Individually ventilated cage (IVC) systems provide cost savings by decreasing the frequency of cage changing (Perkins and Lipman 1996; Reeb and others 1998) and by increasing the number of cages of animals housed per square foot of facility floor space. These systems are increasingly popular and are now widely used.

Of the 63 institutions participating in the 1999 ARS, 30 reported experience with the use of IVCs for laboratory mice. Of those 30, 21 reported that IVCs permitted an extension of the cage-changing interval (Table 11a, Appendix 3). In most cases, this was from twice a week to once a week, but nine institutions were able to achieve an interval of 10-14 days. Thus, IVCs appear to have the ability to reduce labor costs by increasing the total number of cages that a technician can service over a given interval by a factor of 2-3. In a case study provided by Emory University, where changing frequency for a cage of nonbreeding mice went from four times in 2 weeks to once in 2 weeks, the number of cage units serviced per worker per week increased from 780 to 2,000 (personal communication, M.J. Huerkamp). This ratio excludes workers dedicated to cage washing and excludes supervisory personnel. However, an appreciable cost savings in labor, material, and cage replacement resulted that was reflected in lower per diem charges to investigators.

IVCs appear to be suitable for many facility settings and warrant consideration as a method of cost containment in programs that deal with large populations of laboratory mice. The type of contact bedding used in static isolator cages can affect the microenvironment; some bedding types show a significant difference in how long it takes ammonia to reach unacceptable levels (Perkins and Lipman 1995). Use of IVCs and certain types of bedding might allow animal husbandry programs to decrease the cage-changing frequency and still have an acceptable microenvironment. The reduction in labor required to process cages can be significantly reduced and have a major cost saving impact on facilities housing large numbers of rodents.

IVCs can also house many more rodents per square foot of facility space than the traditional method of using shelf racks and standard shoebox cages. This can provide considerable savings in construction costs by reducing the area of the vivarium needed to house a particular number of rodents and the ensuing costs of operating the physical plant (Lipman 1999).

The cost of sanitizing caging and accessories involves more than just wages of personnel who perform the labor. The cage-washing area has inherent liabilities associated with the presence of various chemicals, steam, and conditions that can lead to repetitive-motion injuries of personnel. Introduction of robotics to handle repetitive procedures in the cage-washing area is a recent advance in long-term cost-saving measures. Robotic arms have been designed to process polycarbonate rodent cages through an indexed tunnel washer, working on both dirty and clean sides of the cage-washing apparatus. Robotic technology offers the possibility of substantial long-term cost savings for biomedical research facilities because of its long service life, low maintenance requirements, and the elimination of disability claims in connection with cage washing-related injuries. Whether or not robotics are used for cage washing, automatic dispensers to refill cages with bedding are a useful labor-saving device. As with any investment in labor-saving equipment the institution should compare the labor savings with the cost of the equipment to determine whether the investment is justified.

Newer, more durable polymer plastics are available for rodent cages, with a cost that increases as the strength and durability of the plastic at high temperature increase. High-temperature-resistant plastic rodent cages are superior to standard polycarbonate cages in maintaining transparency and resisting formation of microfissures under conditions of frequent autoclaving (Agee and Swearengen 1995). Facilities that require frequent autoclaving of rodent cages, such as biohazard facilities or rodent barriers, might find that the more durable high-temperature plastic cages, which cost more, would result in savings over time.

Water-Delivery Systems

Although automatic watering systems are a labor-saving device, most mice housed in a variety of cage types in biomedical research facilities are provided water via bottles. When water bottles are used, steps can be taken to maximize efficiency and minimize repetitive-motion injuries associated with manipulating large numbers of the traditional water bottles, sipper tubes, and stoppers. Ergonomically designed tools are available to remove sipper tubes from rubber stoppers and reinsert them later. Water bottles with screw caps or with weep holes (drilled bottles) eliminate the need for rubber stoppers and the effort needed to insert them into bottles, which is considerable. The use of bottle holders with retainer lids that hold several water bottles at once makes dumping and handling of water bottles easier and reduces operation time.

Purchase and Management of Material

Supplies purchased for use in animal care and use programs should comply with the provisions of the *Guide for the Care and Use of Laboratory Animals*. Although supplies account for only a relatively small portion of the budget of an animal facility (about 11% at one major research institution), some cost savings are possible. Here we describe some of the alternative strategies that can be used by animal facility managers to contain supply costs.

Food and bedding are likely to account for a high proportion of the supply costs. Specialty foods can be expensive and should be clearly identified and used only for the purpose defined. Bulk ordering of food and bedding permits obtaining bids on these items and can substantially reduce per-unit cost. However, bulk ordering presents problems for some institutions. Storage of items in bulk requires space, which often must be specifically designed for the items being stored, such as food. Many institutions have multiple animal facilities; in these cases, distribution costs will need to be considered.

Cleaning supplies might also be appropriate for bulk purchase but are subject to the same considerations as food and bedding. Newer facilities have 400-gallon tanks for cage-washing detergent to take advantage of this cost-saving opportunity. Where permitted by facility design and available space, existing facilities might consider retrofitting with equipment that provides greater storage capacity to achieve cost savings. In addition, rack washers are now available with holding tanks that use smaller quantities of chemicals and water.

ENVIRONMENTAL ENRICHMENT

At present, the Animal Welfare Act regulations only mandate environmental enrichment for nonhuman primate species and thus afford institutions the opportunity to contain costs by limiting the application of enrichment strategies to these species. However, there is a growing body of literature on environmental-enrichment strategies for many of the common laboratory animal species, and the *Guide* (NRC 1996a) provides an impetus for institutions to evaluate and incorporate enrichment measures into their animal care and use program for all species where appropriate. Indeed, many biomedical research facilities now provide environmental enrichment to many species of research animals, including rodents. To minimize the resultant increase in the amount of personnel time and facility resources dedicated to these activities, the most labor-efficient devices should be incorporated. For example, if tunnels and other similar devices are used in rodent cages, they should be colorless, nonopaque materials that allow easy visualization of all the animals in the cage. This provision will eliminate the need for additional time and effort to manipulate the devices to permit all the animals in the cage to be seen during observation periods. The 1999 ARS did not provide any information on the magnitude of costs borne by institutions providing environmental enrichment. Eighteen of 52 institutions (see Table 26b, Appendix C) indicated that they subsidized program development costs, such as environmental enrichment, but further details were not given.

A wide variety of enrichment devices and supplies are available as specialty items from commercial sources. However, very good inexpensive alternatives can often be made from other items on hand or from ordinary supplies and materials that are available locally over the counter. Taking that approach potentially carries the dual benefit of involving the animal care staff in a creative, innovative enterprise that contributes to animal well-being and reducing the supply costs associated with this effort. Health and other safety factors should be considered during the design and use of enrichment devices to ensure that neither animals nor personnel are exposed to additional risks.

Animal Acquisition

For animals that are commercially available, inhouse breeding for general animal use is usually more expensive than purchasing animals as they are needed for research studies. Inhouse breeding is required, however, for some studies, such as research on reproductive processes and production of knockout or transgenic animals. About 55% of the mice used in research are purchased from vendors (Table 2). Larger institutions purchase a smaller proportion of mice than smaller institutions presumably because of their more extensive use of transgenic, knockout, or other unique mouse strains in research studies that necessitate inhouse breeding.

Grouping orders can be an effective way to reduce handling and transportation costs. This requires coordination between the principal investigator and the animal facility management to ensure that the animals are available as needed for the research program.

Medical Supplies

Depending on the volume of products used and other institutional circumstances, it might be beneficial to purchase veterinary supplies in bulk or through an institutional pharmacy to achieve cost savings. Drugs and biologics should be stored centrally under appropriately controlled and secure conditions.

TABLE 2 1999 ARS Mean Mouse Census and Proportions of Mice Purchased and Produced in Institutions of Different Size[a]

Institution	No.	Average Daily Census	Purchased	Produced	% Purchased
Group 1	23	9,881	17,426	6,267	74
Group 2	16	19,855	34,722	24,042	59
Group 3	14	46,184	39,233	56,665	41
All	53	22,482	28,199	32,042	47

[a]Table 10a, Appendix C.

Occupational Health

The *Guide* (NRC 1996a) calls for an extensive occupational health and safety program that includes considerable administrative time to establish and maintain the program, to track employees, to train personnel to establish guidelines for the use of personal protective equipment, and to provide for periodic medical evaluation and practice of preventive medicine. Anecdotal evidence suggests that such a program is expensive, but there are no studies of the cost of such programs. This is a subject for further research to ascertain the total cost of such a program and its components so that methods can be devised for cost containment.

Protective clothing and other personal protective equipment, such as gloves, face masks, bonnets, booties, and eye-protective devices, can also be purchased in quantity and provided to staff members as needed. Because the cost of these disposable items can be large, some programs are considering purchasing more durable laboratory coats, jump suits, or coveralls, the most expensive components. In some cases, these items can be repeatedly autoclaved and recycled for reuse to reduce the overall cost. However, the cost of personnel time to collect and autoclave these items needs to be taken into account.

SUMMARY

In summary, the major findings and opinions expressed in this chapter are as follows:

- Animal management, cost accounting, and compliance records are essential for effective management of an animal research facility. They should be kept in a relational database system whenever possible.
- Animal research facilities should carry out cost analysis with such

a method as described in the CARS Manual (NIH 2000). The cost analysis should be examined for areas of potential cost savings and be the basis for setting fees.

• For efficient animal research, an institution can provide core laboratories for a number of services, such as cryopreservation of embryos and semen, monoclonal-antibody production, production of transgenic and gene-knockout animals, histopathologic analysis, and experimental surgery.

• There is a clear economy of scale in research facilities. Labor productivity was markedly greater in institutions with fewer but larger facilities. Institutions should strive to centralize their animal care to as few sites as is compatible with research use.

• Physical plant factors are an important element in the cost of operation of an animal research facility. The physical plant should be designed with efficiency and long-term reliability in mind, and it should be well maintained.

• Individually ventilated caged (IVC) systems provide a satisfactory environment for animals with reduced frequency of cage changing. This results in savings in labor and supplies. Institutions should compare the potential savings from such systems with their cost and invest in IVCs whenever it is justified.

• Automatic watering systems are a labor-saving device. However, if water bottles are used, steps should be taken to maximize the efficiency of the change and filling process, such as use of automatic fillers, use of ergonomically designed tools to remove and reinsert sipper tubes, use of bottles with weep holes, and use of larger bottles to reduce change frequency.

• Supply costs can be reduced through judicious selection of items used and through bulk ordering.

3

Veterinary Medical Care

Veterinary medical care is an essential component of any animal care and use program. The size, scope, and function of the veterinary care program depend on the extent and type of animal care and use. Specific factors that influence the program of veterinary care include the number and type of animal species, the disease backgrounds of animal species maintained, the numbers of animals used, and the experimental characteristics and requirements of the animal models necessary to satisfy research objectives. At a minimum, the veterinary medical care program must be sufficiently robust to satisfy regulatory requirements. Ideally, it is comprehensive and fully integrated into the fabric of the institution, providing demonstrable contributions to the goals of the institution, the research programs, and the overall animal care and use program. In the development of a program of veterinary medical care, there are decision points concerning staffing, sophistication of diagnostic support, and intensity of disease surveillance, which can have considerable cost implications.

Cost effectiveness is an important concern and goal in today's competitive research environment, but quantifying the return on investment in veterinary medical care is difficult. For example, costs associated with a disease outbreak or loss of animals in a specific study could be estimated, but the relationship of the expected frequency of such occurrences to the composition of the veterinary medical program is difficult to assess. Also, relief from the boredom of repetitive tasks is often achieved by rotating assignments among the veterinary care staff, and this further complicates efforts to quantify and analyze cost effectiveness. For example, it is not uncommon for a veterinarian to be responsible for specific research project support, administrative duties, and veterinary medical care responsibilities. Understanding the potential risks (such as disease outbreaks) of a minimal or poorly functioning program is essential to designing a veterinary medical care program that is reasonable and cost-effective.

An assessment of research program needs and regulatory requirements is critical to development of a cost-effective veterinary care program. The assessment should be followed by an effort to design and establish an integrated veterinary medical care program that remains interactive with the research staff and efficient in the delivery of veterinary care while satisfying disparate institutional needs. Making periodic adjustments to the program in an environment of changing research directions and new technologies requires frequent interactions with key personnel in research and administration.

VETERINARY STAFFING

Compensation for professional staff can constitute the greatest operational cost for the veterinary medical care program, and portions of it are often subsidized (Table 23a, Appendix C). Increased numbers of specialty-trained veterinarians are being employed by research institutions as the science and technology of laboratory animal medicine and veterinary medical care advance and their value to research organizations is increasingly recognized. In addition, the growing regulatory burden (NIH 1999) has increased the involvement of specialty-trained veterinarians, particularly laboratory animal veterinarians, in biomedical research institutions. The higher cost of using veterinary specialists has prompted some institutions to look for ways to contain cost through management techniques such as delegation, empowerment, and teamwork to optimize the use of talent. Consultants and part-time employees, both veterinarians and animal care staff, can also be useful in some settings if oversight is adequate to ensure quality and regulatory compliance.

Laboratory animal veterinarians are variously employed by institutions as animal care and use program directors, managers, and clinical veterinarians. Depending on the size and function of the veterinary care program, one or more veterinarians might be needed to satisfy institutional needs. Veterinarians' salaries are higher than those of other veterinary support personnel, so institutions should make use of the veterinarians so as to take full advantage of their professional competences while technical and administrative duties are delegated to lower-paid employees (Gehrke and others 2000). Veterinary residents and certified laboratory animal and veterinary technicians can be used as an effective extension of the veterinary medical staff, as noted in the CIC Study (Appendix B). In addition, in circumstances where a veterinarian is required only part-time, institutions can choose to use consultants, share positions with other institutions, or use the veterinarians' professional competences in research or research-support activities. In the latter case, collaboration between the veterinary staff and the research staff might translate into cost savings for both because a veterinarian would provide skilled assistance while performing required oversight.

Important factors in determining the appropriate level of staffing of veterinarians are the mix of species, the presence or absence of a surgery program, and the use of animal models that require intensive veterinary oversight and assistance because of experimental complications, invasive procedures, or spontaneous disease. Rodent-only programs might require less clinical veterinary support than programs that use larger species or involve animal models entailing surgery or other invasive manipulations that affect animal health and welfare. In the committee's experience, many institutions are finding that transgenic animals require more veterinary support than standard rodent models to deal with breeding issues, health problems associated with unique phenotypes, and the requirement for closely observing the animals for unusual health and animal husbandry problems. In addition, these animals are extensively exchanged among investigators within the country and internationally, increasing the requirement for clinical and diagnostic health assessment programs. Veterinary medical care requirements for surgery-intensive programs include such services as preoperative and postoperative care, diagnostic services, treatment, surgery, and specialized facilities and equipment.

TECHNICIANS

Trained and highly competent technicians are increasingly viewed by institutions as required for efficiently delivering veterinary medical care services in support of higher-paid veterinarians. Many institutions have minimized costs, maximized the use of personnel, and provided valuable career opportunities by delegating responsibility for performing a wide variety of standard veterinary techniques—and advanced research and surgical assistance—to talented and technically proficient veterinary technicians.

DIAGNOSTIC LABORATORY SUPPORT

Clinical pathology laboratory support is a critical component of a high-quality veterinary medical care program. Involving a laboratory-animal-trained veterinary pathologist enhances the quality of such laboratories. Hematology, biochemistry, parasitology, microbiology, and histopathology laboratory services are necessary for disease diagnosis, health surveillance, vendor animal health assessment, and research support. The type and volume of diagnostic laboratory support depend on a variety of factors, including program size, species mix, surgical load, source of animals, and research-support requirements. Institutions must decide, on the basis of cost and quality, whether services should be developed internally, referred to outside contract laboratories, or a combination of the two. Inhouse laboratories are generally more responsive and can be tailored to the species being used. However, startup, staffing, and space costs can be considerable. In contrast, contract laboratories, although not always able to be as responsive as inhouse laboratories, can often deliver services at a lower cost because of economies of scale and a broader testing repertoire. There are some inherent shortcomings in some contract laboratories, including availability in the region of the facility, unfamiliarity with animal specimens or animal diseases, and quality control. Appropriate quality control should be exercised if the results are to yield high-quality research data.

For most small to medium institutions, a combination of minimal inhouse laboratory support with the use of outside contract diagnostic laboratories is most cost-effective. Another option is to share resources among several institutions; this results in cost savings and improves program quality. In large institutions, a dedicated laboratory that is appropriately staffed and equipped might be cost-effective and more responsive. Technologic advances have led to kits for rapid, inexpensive inhouse serologic testing for common rodent viruses and a variety of other assays, allowing smaller institutions to perform some of their own laboratory testing cost-effectively. Many institutions have also found it possible, with little investment, to augment existing research or hospital laboratories and use existing personnel to meet their laboratory animal needs while decreasing overall costs. The optimal approach or combination of approaches can be determined only through careful case-by-case analysis.

HEALTH SURVEILLANCE

An appropriately designed animal health assurance program addresses prevention, control, and treatment of animal disease. The increasingly widespread availability and use of microbiologically defined animal models and the growing recognition of the confounding microbial effects of infections and other diseases has created a substantial demand for health-surveillance programs that monitor the microbiologic status of laboratory animal populations. Also, there is a need to determine the microbiologic status of tumors, cell lines, and other products of animal origin that might be injected into research animals. Cost components include salaries for veterinarians and technicians and laboratory costs for diagnostic and surveillance testing. Researchers are increasingly sharing animals among institutions—animals that could have an unknown health status. This practice has led to an increased need for health surveillance. As animal-housing technology and facility design improve, the maintenance of disease-free, microbiologically defined animals has become a nearly universal standard of care, increasing the importance of disease surveillance (NRC 1996, pg. 27-30).

The planning of health-surveillance programs must include identification of the target populations, definition of program elements, frequency of testing, and methods to be used (NRC 1996, pg. 85-113). Each is evaluated in the context of the species, sources, facility design, and housing conditions; and an approach for each set of circumstances should be determined. Once the target populations are identified and specific program elements—such as vendor surveillance, disease prophylaxis or vaccination, routine observation and reporting, microbiologic monitoring, and histopathologic examination—have been identified, the more difficult task of determining the frequency of testing and the preferred methods must be resolved. Health surveillance is expensive, and many institutions strive to develop a cost-effective program. In particular, the cost of sampling a statistically significant portion of the total population in a surveillance program is often prohibitive. Detecting disease in microbarrier caging systems requires sampling nearly every cage over time by the transfer of bedding to sentinel cages. Consequently, after a careful and informed analysis of risk, staff might opt to reduce costs by lowering the frequency of testing or using less-expensive screening tests initially and then more definitive and more expensive tests as deemed necessary.

SUMMARY

In summary, the major finding and opinions expressed in this chapter are as follows:

- Veterinary medical care programs should be carefully designed to maximize use of the specialist's time by using managers, visiting residents, and certified laboratory animal veterinary technicians.
- The level of veterinary medical care depends on the species mix, size of surgery program, and complexity of animal models used in research.

- Diagnostic laboratory support is a critical component of the veterinary medical care program and can be provided by inhouse laboratories, contract laboratories, or a combination of the two.

- A well-designed health-surveillance program that ensures higher-quality animals is critical to obtaining accurate research results. The surveillance program must be appropriate to needs yet contain costs.

4

Integration of Design, Equipment, Operation, and Staffing: A Contemporary Case Study

The characteristics of physical facilities for housing animals have not changed substantially in the last 10 years. Room sizes, corridor systems, cage and rack systems, finishes, and physical labor have changed little. The ability to genetically alter mice has led to exponential population growth and changes in the physical environment for their care. The impetus for the change is the value of these genetically altered animals, rising operational and per diem costs, and the difficulty in attracting and retaining highly qualified animal care staff. Four of the top 10 medical schools (in terms of grant money) have mouse populations exceeding 25,000 cages and have become mouse research and breeding facilities and yet contain no automation.

With proper facility design, cost-effective care of large mouse colonies and attendant sanitation of cages and racks can be achieved. At the new 55,000-cage mouse facility of Baylor College of Medicine in Houston, Texas, the FY 2000 per diem rate of $0.31/cage (without a filter top) is projected to be reduced when the new $40 million facility is occupied by the middle of 2000. The initial investment is to be recovered from per diem charges. (Note that all prices are in year 2000 dollar amounts and are given for illustrative purposes only; actual prices can vary.) Proper facility design, although requiring a large capital investment, should reduce per diem costs. Many of the lessons learned from designing animal facilities to house 20,000 or more mouse cages cost-effectively can be adapted to smaller facilities. The Baylor College of Medicine project is referenced many times in this section because of the emphasis spent on reducing costs through life-cycle cost analysis, innovation, and adaptation. It should be noted that many of the projected costs and cost savings are estimates made during the design phase. Actual results will be known several years after this project is completed. The detailed analysis is presented here to highlight the necessity for a comprehensive planning process and the need to define goals and set targets. The design team should include the facility director, facility manager, researchers, and representatives of the animal care staff who bring day-to-day front-line experience.

With direct labor representing 50-65% of operating costs, investment in technology that reduces staff or makes current staff more efficient is critical. The committee's recommendations are organized around physical and operational issues.

VENTILATED RACKS

Many institutions have used ventilated microisolator cage and rack systems to extend cage-changing intervals from twice a week to once a week or once every 2 weeks. This extension of the cage-changing interval could allow a doubling of the mouse-cage census without substantially increasing the number of staff involved. Lengthening cage-changing intervals also decreases the load for the cage-wash centers because each cage is washed less frequently. (However, since laboratory animal care technicians also clean rooms, take censuses, receive animals, and support area management, material transport, training, and meetings in addition to cage changing, it should not be expected that halving the cage-changing frequency will lead to a doubling of productivity.) The capital investment in ventilated micro-barrier cages and racks is substantially larger than in static microbarrier housing systems. For example, a 126-cage ventilated rack with water bottles costs 139% more than a double-sided static rack, and a ventilated rack with automatic watering costs 230% more. However, site-by-site comparison of these cage and rack systems, considering total operational costs (equipment, sanitation, personnel, and space), typically indicates, on the basis of committee experience, a payback period of under 5 years for the higher initial investment. Payback periods will vary considerably, depending on the current and projected cage-rack systems, cage-changing frequencies, use of water bottles or automatic watering, mechanical HVAC capacity, room size and configuration, and volume equipment discounts. For some large operations, the payback period is not an important consideration, because hiring and retaining sufficient staff are difficult during a tight labor market. Unless an institution plans to extend the cage-changing frequency substantially (for example, from once a week to once every 2 weeks) or increase the density of cages per rack (from 84/rack to 126/rack or 140/rack, for a 50% or 67% increase), using ventilated racks might not be warranted. Besides the high initial cost of ventilated racks, drawbacks include poor visibility into the cage, the ergonomic stress involved in viewing the bottom and top shelf, and rack weight. Recent modifications, such as rear-mounted feeders and shelf-free rack designs, have improved cage visibility. Ergonomic access can be addressed by assuming 80% and 90% rack use with 140-cage (top and bottom shelf) and 126-cage (bottom shelves) ventilated racks, respectively. With such rack use, the bottom or top row of cages (or both) can be used to temporarily accommodate extra caging or expansion without an increase in floor space. Ventilated racks are heavy—typically 1,000 lb or more when fully loaded. Designing a room where only minimal rack movement is required or increasing the caster diameter from a standard 5 in. to 8 in. can assist with the weight issue. In planning new facilities with ventilated racks and 2-week cage-changing intervals, it should be assumed that 10-20% of the cages will be changed once a week to accommodate special mice strains, such as mice with naturally occurring or experimentally induced diabetes.

VENTILATED-RACK SUPPLY AND EXHAUST

Ventilated racks can be configured with integral HEPA supply and exhaust blowers or connected to a building supply and exhaust. Ventilated racks that do not capture exhaust are not recommended, because heat, allergens, and odors can be returned into the room unless the exhaust is HEPA-filtered. Institutions using large ventilated racks can profit from direct connection to a building HEPA supply and be nonfiltered (or filtered, depending on location and application) because of cost savings, ventilation redundancy, and lower maintenance costs. At Baylor College of Medicine's new facility, the decision to build a HEPA-filtered building supply system instead of using individual rack systems saved over $16,000/room (supply and exhaust HEPA blowers would cost $2,500/rack, and each room has nine racks, for a cost of $22,500/room; but building supply, exhaust, and ductwork cost only $6,500/room). With individual rack systems, if the blower fails, ventilation rates revert to a static state. Using building systems with redundant supply and exhaust units on emergency power allows uninterrupted ventilation to each rack. For more information on ventilated racks, see Lipman (1993).

AUTOMATIC WATERING

A 16-oz water bottle in a microisolator with four to five mice in it will not be sufficient for 2 weeks. Extending cage-changing frequencies to once every 2 weeks requires automatic watering, weekly changing of the water bottle, or a larger water bottle (28-30 oz). Many institutions using ventilated racks with a 2-week cage-changing frequency use automatic watering. Early automatic watering systems with the valve attached to the cage were prone to leaks or mouse dehydration because of improper docking of the valve. On the basis of committee experience, recent automatic watering systems with the valve attached to the cage, if docked appropriately, perform as well as water bottles. Replacing cages on the rack requires priming of the valve by cage-changing personnel and researchers. Automatic watering systems with the valve attached to the rack do not require priming but should be wiped with a disinfectant before cage replacement to prevent cross contamination. Changing standard 16-oz water bottles weekly and cages every 2 weeks might be practical, especially where the water bottle is outside the cage. At Baylor College of Medicine, investigators' rejection of automatic watering necessitated redesign of the low-profile microisolator top to accommodate a 28-oz water bottle and a 2-week cage-changing frequency. Baylor conducted clinical trials by acidifying the water to a pH of 2.3 and confirmed that the 28-oz water bottle did not exhibit bacterial or fungal growth in 14 days (Robert Faith, personal communication). Water bottles pose serious labor and ergonomic issues for an animal facility. Uncapping, washing, filling, recapping, and sterilization are time consuming and labor intensive and can lead to repetitive-motion injury.

UNIVERSAL ROOM DESIGN

An animal housing and research room (AHRR) size of 16×22 ft can accommodate a wide variety of racks, pens, and species. Mouse AHRRs with an average of two mouse cages per assignable square foot (ASF) are considered to have high density (Table 3).

At Baylor College of Medicine, researchers rejected the typical six double-sided mouse racks arranged library-style because only 3 feet was left between the faces of racks, necessitating movement of the 1,200-lb

TABLE 3 Number of Cages per Square Foot by Percent Rack Use[a]

Fraction of Racks Used, %	Total No. Cages	No. Cages per Square Foot
80	672	1.91
90	756	2.15
100	840	2.39

[a]Assuming 16×22-ft or 352-ft^2, room with six 140-cage racks.

ventilated racks during cage-changing and procedures on the animals. Breeding rates in some mouse strains were reduced when racks were moved (Robert Faith, personal communication). In response to those issues, the room was configured with three single-sided racks against each 22-ft wall and three double-sided racks down the middle. The six single-sided racks and three double-sided racks yielded the equivalent of six double-sided racks with 5-ft between the faces of racks. If the 5-ft aisle is used as procedure space and cage-changing space, the ventilated racks are only moved two to four times per year for washing. During cage-changing, an animal transfer station is moved down the 5-ft aisle, bringing the transfer station to the cage, in contrast with what happens with the library style configuration, in which cages are brought to the transfer station. Single-sided ventilated racks cost 75% as much as double-sided racks, and the drawback to this design is higher equipment costs. In the Baylor College of Medicine project, this rack arrangement resulted in an increased cost of equipment of about $18,000/room, or a total increase of about $1 million. This increase was thought justified because it makes the room much more user friendly to research staff and animal husbandry staff. The increased efficiency and reduction of injuries resulting from not requiring movement of heavy racks for cage changing or experimental manipulation of animals will quickly pay back the additional cost. From the 1999 ARS survey, the average for cage-changing per person for group 2 and 3 institutions ranges from roughly 400 cages per week for individually ventilated cages to 800-950 cages per week for other types of caging (see Table 8l-n, Appendix C). Most institutions used a change station for microisolator cages and for individually ventilated cages. Baylor College of Medicine expects at least 300 cages/day per person (roughly 1,500/week) with the revised rack layout and new transfer-station design, resulting in 20-50% increase in productivity per cage changer. Experience will test that expectation and will reveal any ergonomic problems that arise. Room mockups were useful in choosing the final room size and layout.

ANIMAL TRANSFER STATIONS

Transfer stations may be clean-air workstations or biologic safety cabinets with a 10-in. or 12-in. sash opening on one side. The restricted sash opening affects cage-changing frequency and has historically limited cage changes to 200-250 cages/day per person. Some institutions use workstations with two sash openings(front and back. At Baylor College of Medicine, a new four-side open transfer station was developed to take advantage of the 5-ft aisle between racks and to increase cage-changing productivity to 300 cages/day per person. The advantages of the new transfer station include an adjustable 18- to 24-in.-high sash, allowing unencumbered hand movement, and a team approach to cage changing. The new four-sided transfer station is a workstation and does not have the biologic-containment properties of a biologic safety cabinet, so only product protection is provided.

ROBOTICS

For 4 years, three animal facilities in Sweden have been successfully operating numerous cage-washing facilities with robots handling the monotonous and repetitive chores of dumping waste from cages, placing cage components (bottom, top, wire bar lid, and bottle) on tunnel washers, removing cage components from tunnel washers, and filling cages with bedding. The principal motivation for using robotics in Sweden is the recognition that the highest percentage of work-related injuries in an animal facility occur in the cage-washing area because of repetitive-motion injuries, sensitization to allergens created during cage-dumping, and heavy lifting. To comply with occupational health and safety rules in Sweden, which require proof that a task associated with health hazards can be performed only by humans, directors of animal research facilities have explored the use of robotics. The cage-washing area typically experiences the highest staff turnover rate. The potential of robotics to decrease costs remains to be determined. At Baylor College of Medicine, robots will process the 55,000 soiled and clean cages per 2 weeks (10 working days) with indexing tunnel washers (a tunnel washer that moves a batch of cages at a time through the various (prewash, wash, rinse, and dry) treatment compartments), conveyors, and a vacuum bedding system. In a presentation to Tradelines, a for-profit seminar group, data provided by Baylor College of Medicine indicated that the $1.2 million premium for using robots, indexing tunnel washers, a vacuum bedding system, and special material-handling equipment resulted in a payback of 4.11 years. The robots have been successfully used in many automated production facilities in the automotive industry for over 20 years with a mean time between failures of 50,000 hours for the entire robot assembly. Robots should be seriously considered for facilities that process 4,000-5,000 cages/day (four staff at two tunnel washers) and evaluated when cage-processing reaches 2,000-2,500 /day (two staff at one tunnel washer). The cost of cage-processing robots is expected to decrease as more installations come on line and engineering costs are amortized over many projects. With a projected growth of 20-22% per/year in mouse census, robots will allow animal facilities to redirect valuable staff to animal-husbandry functions rather than monotonous and repetitive cage-washing activities.

VACUUM BEDDING SYSTEM

Handling of soiled and clean bedding in an animal facility is a labor-intensive task. Soiled bedding is removed from cages and waste is hauled to a dumpster manually at most animal facilities. Clean bedding can be automatically dispensed at the end of tunnel washers by manually filling hoppers of an automatic bedding dispenser from 40 to 50-lb bedding bags. Vacuum bedding systems can be used manually or in conjunction with robots to pneumatically transport soiled bedding to remote dumpsters and transport clean bedding to bedding dispensers. The vacuum creates a downdraft at the dump station, minimizing environmental dust and allergens. There are two other waste-disposal systems. One grinds up the waste and bedding, adds water, moves the waste by a pipe to a press that squeezes out the water, and puts the waste in a dumpster. A related method is to grind up the waste, add water, and discharge into the sewer. One must check with local authorities to use this method.

EXPANDABLE-CONTRACTIBLE BARRIERS

Most animal facilities are designed with a fixed percentage of barrier space (housing space that isolates animals from contamination) (NRC 1996, pg. 119). Although conventional and barrier-space entry protocols for people, animals, and materials vary with the institution, for purposes of this report, a barrier will be defined as personnel fully gowned (gown, booties, gloves, face mask, and cap) and all material (racks, cages, feed, and bedding) autoclaved before entry into the barrier. Because animals housed in a barrier often have higher per diem costs to reflect their special treatment, underuse of a barrier facility or use of a barrier facility to house conventional animals can increase operating costs. Designing an animal facility with an expandable-contractible barrier can be cost-effective if a single-directional corridor system with multiple doors or air locks is used, beginning at sterile-equipment holding and terminating at soiled-cage washing. Alternative emergency exits must be available. By using this concept, the barrier can be sized from an individual room or suite up to the entire facility in selected increments.

INTERSTITIAL SPACE

Interstitial space is defined as an accessible zone that permits personnel movement above the ceiling of a facility and is typically used for maintenance or modification of HVAC equipment and utilities serving the space below. Animal facilities are mechanically complex and require constant maintenance. Easy access to terminal reheat coils, dampers, ventilation ducts, utilities, shutoff valves and such HVAC equipment as HEPA filters (if used), and supply and exhaust boxes is critical for the proper operation of an animal facility. Most animal facilities are serviced from within the facility through access panels or lay-in hung-ceiling assemblies that require a 14- to 16-ft floor-to-floor height. A partial interstitial or full interstitial space above an animal facility is desirable and sometimes essential to maintain a barrier or containment facility, eliminate the need for access panels or lay-in hung ceilings, restrict personnel access, reduce noise, and perform routine or emergency maintenance. Proper design of interstitial space carefully coordinates the placement of all ventilation ducts and utilities while maintaining unobstructed service aisles. Partial interstitial space provides a walk surface above a part of the facility—typically over corridors—and requires a 16- to 18-ft floor-to-floor height. Full interstitial space provides a walk surface above the entire facility and requires an 18- to 20-ft floor-to-floor height. The increased cost of constructing an interstitial space over that of conventional construction is related to the greater floor-to-floor height (deeper basement or more exterior wall, depending on the animal facility location), the walk surface, and the mechanical coordination needed to create service aisles. The exact increase in costs will vary from one project to another and should be estimated accordingly. On a recent two-level, 103,600-gross-square-foot animal research facility project, the cost of partial interstitial space was $705,000 (catwalk, $345,000; excavation and structure, $195,000; and mechanical, $165,000) and for full interstitial space, $2,665,000 (additional floor, $560,000; excavation and structure, $1,650,000; mechanical, $455,000). The increase in costs can be offset by a lower life-cycle cost achieved through ease of access for maintenance over the life of the facility, with some initial savings realized during construction because multiple trades can work simultaneously above and below the ceiling.

WALL MATERIALS AND FINISHES

Concrete masonry units (CMUs) have been used extensively in animal facility construction because of their durability and familiarity. The quality of CMU installations can vary considerably, depending on the surface quality of the block, the dimensional stability of the block, installation, filler application, primer, and final paint coats. Typically, wall guards are added to protect the painted finish as well. Other materials—such as water-resistant gypsum wall board (WRGWB), solid cement board (Titon-Board®), and fiberglass-reinforced panels (FRPs)—have been used successfully in rodent-based animal facilities. Titon-Board is a unique product consisting of solid cement board with a smooth face. The relative costs of these installed wall systems are as follows: CMU with epoxy paint, $19/ft²; Titon-Board with epoxy paint, $11/ft²; 4-mm FRPs, $25/ft²;

6-mm FRPs, $27/ft^2; and WRGWB, $9.50/ft^2. The board-panel assemblies can be constructed quickly and result in a very smooth finish, compared with CMUs; with wall protection, they can hold up well against the demands of rodent-based animal facilities.

SUMMARY

In summary the major finding and opinions expressed in this chapter are as follows:

• Proper design of the animal facility is a major determinant of the institution's ability to deliver cost-effective animal care. The design team should include the facility director, facility manager, researchers and representatives of the animal care staff with day-to-day experience in the facility.

• Cost reductions should be calculated over the life the facility and take into account equipment, material and workforce interactions and durability.

• Labor savings are a distinct advantage for the use of ventilated rack systems for mice due to the reduction in the frequency of cage changing. In addition, ventilated cage systems connected to the room exhaust have the advantage of improving room air quality and reducing worker exposure. Careful selection and analysis of available ventilated cage systems for the conditions of intended use are necessary for a sound financial decision and improved operational efficiency.

• The use of conventional water delivery via bottle is laborious, time-consuming and likely to produce repetitive motion injuries in personnel. Automatic watering systems and alternative water bottle design and methods of handling warrant evaluation as a possible cost-saving, injury-sparing measure.

• Designers of animal rooms should take into consideration ease of equipment use and animal handling to reduce worker fatigue and injury.

• The use of robotic equipment to perform monotonous tasks, such as preparing cages for washing, is projected to have financial advantages and to reduce the incidence of ergonomic injuries in personnel. Robotic equipment may prove to be a viable investment for institutions processing as few as 2,000-2,500 cages daily.

• Interstitial space for access to the animal facility mechanical areas should be provided because these areas require frequent preventive maintenance and repair services that are disruptive to ongoing research and smooth facility operations.

• Wall materials that are durable but less expensive than the widely used concrete masonry units may be appropriate in some animal facility applications.

5

Regulatory Concerns

The Institutional Animal Care and Use Committee (IACUC) plays a critical role in an institution through review and approval of research protocols and semiannual review of the institution's facilities and programs for the humane care and use of laboratory animals. It is important to note the interactive relationships of the IACUC and the animal research program in the assurance of high-quality care. The IACUC has responsibility for oversight of all components of laboratory animal management, so poorly managed or chronically undersupported animal research facilities and programs not only erode the research mission and cooperation of investigators, but also require an extraordinary commitment of time and effort on the part of the IACUC. Ill-advised reduction in support of research animal program administration could result in a degradation of the program and increased expenditures related to regulation. Personal communications from several financial officers at academic institutions have indicated that the magnitude of IACUC costs is underestimated by many institutions because the institution fails to account for the cost of faculty time spent on IACUC activities. For those reasons, most institutions rely on strong leadership of the animal care and use program to diminish costs of IACUC program oversight

Institutions acknowledge the importance of maintaining viable regulatory compliance, but researchers and administrators at universities have complained for many years about the high cost and time required to comply with federal and state regulation of the use of animals in research. However, compliance cost has been difficult to estimate. In 1995, seven major research universities tried to estimate the cost of complying with pertinent federal regulations (Greger 1995). The University of Wisconsin-Madison in 1995 employed eight full–time equivalents (FTEs) to support the efforts of college and all-campus IACUCs. These people (including veterinarians part of the time) processed protocols, attended IACUC meetings, performed animal facility site visits, and educated faculty, other researchers, and IACUC members on animal care and federal compliance issues. Faculty serving on IACUCs contributed the equivalent of 4,000 hours/year (2 FTEs) in reviewing protocols, attending IACUC meetings, and participating in semiannual facility inspections. On the average, the faculty and staff spent 19 hours per protocol to meet compliance recommendations.

No attempt was made to estimate the amount of time that investigators spent in preparing and revising protocols. The costs of animal care staff, veterinarians, and institutional review board members to attend national and regional training was not estimated. Future surveys should gather information regarding these costs as an overall assessment of training costs.

All seven universities agreed that the amount of faculty and staff time spent on compliance with animal use regulations was large and did not necessarily reflect the quality of animal care programs. Some types of protocol took more time to review—those involving international collaborators, those with complex and multiple procedures, and especially "less developed" protocols. Accordingly, the seven institutions recommended "just-in-time" review of human and animal use protocols with no review of protocols submitted to the National Institutes of Health (NIH) and "considered unfundable" by a study section. Depending on the institution, this would eliminate the need to review 10 to 50% of protocols submitted for NIH funding. However, this recommendation would not reduce the IACUC's workload for proposals submitted to industry, the National Science Foundation, or the US Department of Agriculture (USDA), because their grant review differs from that of NIH.

During the next 3 years, the so-called regulatory burden was often mentioned but never analyzed successfully. However, the House of Representatives Committee on Appropriations (House Report 105–205, p. 98) in the FY 1998 budget report mandated that NIH conduct a study of regulatory burden. The mandate extended the study to "regulations governing use of animal and human subjects in research and regulations covering the use and disposal of hazardous and radioactive materials." NIH convened a focus group of researchers, IACUC members and staff, and laboratory animal veterinarians to assess animal care and use issues. The resulting report (NIH 1999) cited the following as major categories of problems:

- Redundancy of program review and inspections.
- Inconsistency in yearly reports required by the Office for Protection from Research Risks (OPRR), USDA, and the Association for Assessment and Accreditation of Laboratory Animal Care International (AAALAC).
- Inconsistency between USDA and OPRR on protocol review.
- Outdated or poorly conceived USDA requirements, including those dealing with caging of animals.
- Inconsistency in interpretation of regulation and policies by oversight groups.
- Complexity of regulations governing the transportation of animals and materials derived from nonhuman primates.

No estimate was made of the cost of complying with the redundant or inconsistent policies. However, the Animal Care and Use Workgroup noted that the Joint Commission on Accreditation of Healthcare Organizations (JCAHO) accreditation review of hospitals occurs every 3 years. In contrast, "IACUCs are required by the Health Research Extension Act and the Animal Welfare Act to conduct both an in-depth review of the institution's program for the humane care and use of animals, and an inspection of its facilities every 6 months. In addition, current law also requires that the USDA inspect every facility once a year. Furthermore AAALAC conducts a full accreditation site visit every 3 years for those institutions that voluntarily seek accreditation" (NIH 1999). As a result the animal care programs and facilities at an institution are reviewed at least 3 times per year. Some experts in the regulatory work group think that reducing redundancy and inconsistency of efforts would allow faculty and staff to spend their time more efficiently in producing high–quality research with well-tended research animals.

This committee is not aware of studies documenting the costs of training investigators in writing protocols and training required before procedures are performed as well as the costs of training research staff in record keeping and the proper use of animals. However, if properly done, these training costs must be considerable. A potential benefit would be that well-trained staff perform more efficiently.

The 1999 ARS demonstrated that the costs of supporting IACUC functions are substantial, even apart from faculty time, and are a frequent recipient of institutional subsidy. Of 48 institutions that responded, 31 reported that their IACUCs had an annual budget in excess of $50,000 (range, $0–301,260) (Table 29, Appendix C); and in 27 of 51 institutions that responded, the IACUC budget was funded in whole or in part by the institution (Table 26b, Appendix C).

The 1999 ARS was not designed to address the regulatory burden issue, but it yielded some insights into the topic, especially with regard to the relative burden for small and large research programs. The 14 institutions with large animal use programs (group 3) invested more in the management of regulatory compliance than the 23 institutions with smaller programs (group 1) (Table 29, Appendix C). They were more apt to have a program for monitoring animal experimentation apart from the mandated semiannual IACUC inspections (92% versus 70%), had more faculty and staff serving on IACUCs (21 versus 14 members), and budgeted more for IACUCs ($164,000/year versus $63,000/year). However, the cost of compliance as a percentage of research dollars received was generally higher for smaller programs (Tables 21b and 29, Appendix C).

The proposal to require USDA to regulate the use of rats, birds, and mice in research—as well as other species—will increase the regulatory burden in all institutions. However, the burden will be especially heavy in smaller institutions that have had no previous regulatory experience and in institutions that depend on difficult to obtain state funds and state approval for renovation of facilities.

The 1999 ARS also provided insights into the issues that most concerned laboratory animal veterinarians and users of research animal facilities. They ranked their concerns in descending order as high per diem rates, inadequacy of space available for animal housing, and burdensome regulatory compliance and inadequate institutional support for the facility (tied) (Table 31, Appendix C).

Some noted that the large investment that institutions must make to support regulatory compliance reduces the funds available for renovation and expansion of animal facilities or reduction of per diem rates. Although that might not be true in all cases, 13 of 52 researchers perceived institutional funding of animal research to be inadequate.

Perhaps the biggest shortfall is in funds for upgrading of animal facilities. This is due to at least 4 factors:

• Most research institutions have delayed maintenance of their research facilities. Thus, funds for renovations are used for both repairs and upgrade.

• Transgenic animals and modern research techniques require ever larger and more sophisticated animal facilities.

• NCRR has a small budget for upgrading animal facilities. NIH Research and Program (R and P) series grants provide little support for facility renovation.

• Universities are relying more on donations for facility upgrades, but animal facilities are less appealing to donors than other facilities (partially because of the activities of animal-rights activists.)

The result of the costs of complying with regulations is that institutions and researchers have tried to become more efficient in all aspects of animal research. Most experts think that reducing the regulatory burden on animal use is one way to make animal care more efficient.

SUMMARY

In summary, the major findings and opinions expressed in this chapter are as follows:

- Costs of regulatory compliance are usually underestimated because costs of faculty time for IACUC activities, and for writing protocols as well as costs of training are rarely assessed. Just-in-time protocol review might reduce costs somewhat.
- Some regulations governing use of animals in research are redundant and inconsistent; this leads to increased costs.
- The IACUC annual budget was greater than $50,000 for 65% of institutions responding to the 1999 ARS survey. The budget was somewhat higher, when calculated as a percentage of animal research dollars, for smaller institutions.
- The proposal to require USDA to regulate rats, mice, and birds will be especially burdensome for smaller institutions. This and previous items would suggest forming independent IACUCs to handle the compliance needs of smaller institutions.

6

Future Directions in
Research Animal Use:
Infrastructure, Cost, and Productivity

OVERVIEW

The information in this chapter is based on the experience of the committee members and informal consultation with a number of investigators who use a considerable number of animals in their research. The purpose is to facilitate planning by projecting the likely expansion in the use of animals. Data contained in the US Department of Agriculture's annual Animal Welfare Report show a decline in the use of all animals covered by the Animal Welfare Regulations over the last decade, from 1.75 million in 1989 to 1.2 million in 1998. The use of all species except nonhuman primates fell. However, rats, mice, birds, and all cold-blooded animals are excluded from coverage. It is estimated that over 90% of animals used in research are mice and rats. It seemed important to examine trends in the use of mice because it is the committee experience that such use will drive the need for new or renovated animal research facilities in the near future.

The major increase in animal research in the last few decades has involved the use of the mouse as an experimental animal. It is likely that the largest increase in demand for animal care will be for mice, although other experimental systems—such as flies, worms, fish, frogs, and pigs—are being further developed and used.

A number of factors influence the use of the mouse as an experimental system. A major initial factor was the development of transgenic mouse technologies in the middle 1980s. Use of transgenes to achieve deregulated or tissue-specific expression of desired genes in mice was an important component of research that led to major breakthroughs in several fields of biology. In cancer research, expression of dominant oncogenes as trans-genes led to the development of basic and applied models for the study of a wide variety of neoplasms. The ability to achieve specific transgene expression has led to a large increase in newly generated mouse models and has resulted in a quantum leap in our level of understanding of the development and function of the immune system.

The first successful application of embryonic stem cell (ES cell)-based approaches to introduce gene-targeted mutations into mice was reported less than 10 years ago (Capecchi 1989). This technology has had an even more dramatic impact than transgenesis on basic and applied research, further establishing the mouse as a major experimental model system. Until several years ago, application of gene-targeted mutation technologies in mice was limited largely to a handful of major research centers or specialized investigators. However, as with most technologies, gene-targeted mutation approaches and reagents have been refined to the point where they are now accessible to most research institutions and are readily used by much of the biomedical research community. This technology for defining mammalian gene function in a physiologic setting, unimaginable 20 years ago, has become one of the most widely applied and most informative tools of biologic research.

Application of gene-targeted mutational analyses is likely to continue to increase demand for the mouse as a model system in the next decade, especially when coupled with powerful new technologies—such as genomics—and the potential power of combinatorial studies of existing or future targeted mutations.

FACTORS CONTRIBUTING TO INCREASED MOUSE USE

The Genome Project and functional genomics, including gene-mapping experiments and gene-function validation, are major factors that will increase the use of mice . The project has rapidly increased the volume of known genetic sequences and identified genes, a large proportion of which have unknown functions. These sequences are being made available in easily accessible genomic databases—leading to more target sequences for gene-targeted mutation. The use of mice for large-scale gene mapping experiments and functional genomics will increase dramatically as these mutagenesis projects get under way. The largest increase in animal use will presumably occur mainly in a small number of large centers and in industry, but the overall impact will be widespread.

Gene identification will become progressively easier as better mouse genetic maps are constructed, although this will lag a few years behind human maps. In the meantime, as the human map nears completion, information from syntenic regions in the mouse might be useful and speed up gene identification based on mapping information. That in turn will lead to use of animals for validation of function. Increased numbers of genes identified through the Genome Project could potentially lead to thousands of new gene-targeting experiments, provided that resources continue to grow.

Developing technologies, such as array analysis, will increase the utility of mouse models. These powerful diagnostic techniques will enable analysis of expression patterns in, for example, tumor models that express a variety of genes in the same pathway. As techniques become more sophisticated, it will be possible to look at early disease stages and to dissect complex interactions in tissues. In addition, gene chips and protein chemistry will require an increased number of animals to generate proteins for analyses.

There are increased interinstitutional transfers of novel lines coupled with combinatorial interbreeding of different lines that will lead to increased use of mice. As an example, some 300 mutant lines were brought into the Dana-Farber Cancer Institute (DFCI) and 300 other lines were sent out of the DFCI in the last year. The ready transfer of lines, coupled with interbreeding of mutant and transgenic lines to generate large numbers of new lines, will result in a large increase in the number of mice used. Examples of the application of interbreeding of lines include:

• Generation of animals with polygenic mutations, using multiple mutant or transgenic backgrounds for basic studies in such fields as cancer biology, immunology, and neurobiology. The combinatorial breeding of different mutant backgrounds could generate huge increases in numbers of experimental mice.

• Genetic-modifier studies, for example, analyses of favorable and adverse influences of genetic background on current or future cancer-model strains.

• Polygenic disease models involving multiple contributing genetic loci with respect to such diseases as cancer and some immune diseases.

• Back-crossing and inbreeding to create the desired genetic backgrounds for immunology studies.

• Conditional mutagenesis.

Conditional targeted mutations and tissue-specific mutations (tet, cre/lox, and other similar strategies) will further increase animal use for modeling and developmental studies. The technology is still being developed, and it will be a few years before it sees widespread use. Rapid improvements could occur if National Institutes of Health (NIH) or foundation resources are targeted to improving and distributing this technology.

Chemical and viral mutagenesis of mouse germline will be used to study environmental mutagenic effects, to identify new genes involved in development and cancer, and to create models for therapeutic trials. Interest in transgenic mice has the potential to increase dramatically. In many instances, well-designed transgenic experiments, potentially in combination with knockouts, can be more informative. Therapeutic models—for example, for cancer therapy, gene therapy of genetic diseases—are expected to increase.

As basic understanding of molecular biology increases, there will be an increasing interest in and emphasis on whole-animal in vivo experimentation. This will increase the use of mice for experiments involving gene transfer into preimplantation and postimplantation embryos and observations of the effects in organ culture and in utero.

The ease of mouse-genome manipulation resulting from the establishment of core laboratories for generation of mutant lines, histopathologic analyses, genotyping, and other analyses will benefit the national genomics initiative if creating these core laboratories becomes a national priority.

An increase in NIH monetary support for infrastructure development and the payment of direct costs could determine the level of animal use. Many institutions are pursuing the construction of new animal space and space renovation for modernization. If the national economy stays robust, the NIH budget should grow and make resources available to continue expanding mouse work. Growth of the infrastructure portion of the National Center for Research Resources budget of the NIH has not kept pace with the need for new animal research space.

New design concepts and technologies are resulting in more efficient and larger animal facilities, which have greater capacity. Many institutions now regard the capacity of their animal facilities as the major factor that limits the expansion of their biomedical research programs.

POTENTIAL STRATEGIES TO DAMPEN THE EXPLOSION IN MOUSE USE

Because of the advances noted above in the use of the mouse as a primary model system for the investigation of mammalian genetics, it is inevitable that the number of mice used in institutional research programs will continue to surge. On the basis of the committee's experience, several useful strategies are suggested to manage growth of mouse populations:

- The use of prudent colony management—especially involving breeding animals, effective database management, and accelerated genotyping—can reduce generation and retention of extraneous animals. These colony-management techniques could stimulate other choices for institutions that might choose to use external specialists with relevant expertise to establish training programs to address specific needs.

- Preservation of lines, embryo freezing, sperm cryopreservation (the least expensive method, pending resolution of issues related to pathogen transmission and long-term viability), and viable in vitro fertilization methods might reduce the need to maintain various mouse mutants as active populations in facilities.

- The use of satellite or centralized animal research facilities might reduce the overall impact on an institution's resources if there are financial incentives to house off-site in commercial contract sites.

- More central repositories for unique mutants are created to meet the higher demand for mutants.

- Alternative central animal research facilities are created through regional consortia or independent academic medical centers with outstanding histories of laboratory animal management.

- Improved animal research facilities are provided that can result in better health of strains and less need for strain re-derivation or regeneration after disease outbreaks or other cataclysmic events.

- Centralized cores for common strains, such as cre/lox and RAG, might reduce overall numbers as investigators become confident about timely strain availability and effective strain distribution.

- In some areas, the mouse might be replaced in genetic studies with simpler organisms that have sufficient homology (such as yeast, *Drosophila*, and *Caenorhabditis elegans*) as a result of genome-sequence determination, but this effect is probably transitory.

SUMMARY OF MOUSE PROJECTIONS

Barring a major decrease in funding, factors that support a substantial increase in use of mice greatly outweigh factors that would decrease their use. Many institutions have projected a threefold increase over 5 years, assuming that space and funding are adequate, but some suggest that such a projection is very conservative. Lower estimates from other institutions (including Harvard and Albert Einstein) might reflect the constraints on space that these institutions encounter.

POTENTIAL FOR USE OF OTHER TRANSGENIC SPECIES

Rat and Rabbit

Technologies have been developed for generation of transgenic rats and rabbits. The use of transgenic rats and rabbits also occurs in academic settings, although this will depend even more heavily on funding because such models are potentially very expensive. Support for these models will depend to some extent on the technologic ability to make physiologic measurements or conduct disease interventions in these animals that cannot be carried out in mouse models.

Pig

Transgenic pigs are more attractive than mice for modeling human vascular diseases and, potentially, organ transplantation. The use of this animal model system in translational research is substantial in the academic setting.

Other Transgenic Mammals and Birds

The application of transgenics or gene-targeted mutations in other large animals or birds could also increase but would probably find most current use in applied science in commercial settings. With the exception of pigs and nonhuman primates, there is no obvious reason to expect an increased demand for large animals in research over the next 5 years.

Xenopus

Some growth in use of *Xenopus* is expected. NIH is considering a plan to initiate a genome project for frogs that involves expressed-sequence tags, using *Xenopus tropicalis* for frog genetics. Frogs have been used traditionally for developmental and cell biology studies.

Zebrafish

The use of zebrafish as a model for studying development has shown a high degree of promise. Zebrafish require relatively low maintenance. Large-scale mutagenesis screens for recessive traits have been successfully carried through to identification and cloning of mutant genes. Those chemical mutagenesis screens have been successful in isolating zebrafish lines that contain mutations affecting organogenesis and neurogenesis, physiologic function of such organs as the heart and a variety of mutations affecting different stages of embryonic hematopoiesis. Most of these mutants live well beyond the stages of early development and so allow identification, propagation, and genetic characterization.

The use of zebrafish for the study of vertebrate embryonic development, neurogenesis, organogenesis, medically relevant pathophysiology, and fundamental mechanisms of cancer might increase exponentially over the next decade. Over the last year, an NIH-sponsored zebrafish genome initiative has been launched and has resulted in a vast improvement in knowledge of the genome of this organism. Large regions of synteny have been identified in the mouse and the human; this indicates that advances in genomic sequencing in these species will also facilitate use of the zebrafish model.

SUMMARY

In summary, the major findings and opinions expressed in this chapter are as follows:

• The Human Genome Project and functional genomics supported by a diverse array of experimental approaches will continue to fuel the use of the mouse as the primary experimental model system in the investigation of mammalian genetics.

• Many strategies may prove to be useful to hedge the ongoing explosion in mouse use. These include: improved colony management; database management; techniques to maintain genetic stocks without maintaining active populations; consolidation of key mutant lines or strains into fewer facilities to eliminate redundant production while maintaining prompt distribution; and continued animal health improvements and the replacement of mice with simpler organisms when applicable.

References

Agee, W. B., and J. R. Swearengen. 1995. A durability comparison of regular, high-tempera-ture, and ultra high-temperature polycarbonate rodent cages under barrier conditions. Contemp. Top Lab Anim Sci 34(3):54. (Abstract)

Capecchi, Mario R. 1989. Altering the genome by homologous recombination. Science 244:1288-1292.

Gehrke, B. C., B. J. Weigler, and M. M. Slattum. 2000. Professional income of laboratory animal veterinarians predicted by multiple regression analysis. J Am Vet Med Assoc 216:852-858.

Greger, J. 1995. Just in Time Report to NRC Round Table, December 7, 1995. Washington, DC: National Academy Press.

Houghtling, J. L. 1998. Outsourcing your animal care program. Lab Anim 27(9):35-39.

Lipman, N. S. 1999. Isolator rodent caging systems (state of the art): A critical review. Contemp. Top Lab Anim Sci 38(1):9-17.

Lipman, N. S. 1993. Strategies for Architectural Integration of Ventilated Caging Systems. Contemp Top Lab Anim Sci 32(1):7-10.

NIH [National Institutes of Health]. 1999. NIH Initiative to Reduce Regulatory Burden: Identification of Issues and Potential Solutions, edited by J.D. Mahoney. http://grants.nih.gov/grants/policy/regulatoryburden/index.htm.

NIH [National Institutes of Health]. 2000. Cost Analysis and Rate Setting Manual for Ani-mal Research Facilities. NIH Committee on Revision of Cost Analysis and Rate Setting Manual for Animal Research Facilities. NIH Publication No 00-2006. Bethesda, MD:US Department of Health and Human Services, National Institutes of Health, National Center for Research Resources.

NRC [National Research Council]. 1998. Approaches to Cost Recovery for Animal Research: Implications for Science, Animals, Research Competitiveness, and Regulatory Compli-ance. Washington, DC: National Academy Press.

NRC [National Research Council]. 1996a. Guide for the Care and Use of Laboratory Ani-mals. 7th ed. Washington, DC: National Academy Press.

NRC [National Research Council]. 1996b. Rodents: Laboratory Animal Management. Washington, DC: National Academy Press. P 167.

Perkins, S. E., and N. S. Lipman. 1995. Characterization and qualification of microenvironmental contaminants in isolator cages with a variety of contact beddings. Contemp Top Lab Anim Sci 34(3): 93-98.

Perkins, S. E., and N. S. Lipman. 1996. Evaluation of microenvironmental conditions and noise generation in three individually ventilated rodent caging systems and static isolator cages. Contemp Top Lab Anim Sci 35(2): 61-65.

Reeb, C. K., R. B. Jones, D. W. Bearg, H. Bedigian, D. D. Myers, and B. Paigen. 1998. Microenvironment in ventilated animal cages with differing ventilation rates, mice populations, and frequency of bedding changes. Contemp Top Lab Anim Sci 37(2): 43-49.

APPENDIX
A

Office of Grants and Acquisition Management Memorandum

Office of Grants and Acquisition Management memorandum concerning the treatment of the facilities and administrative costs of animal research facilities in OMB Circulars A–21, A–122 and Appendix E, 45 CFR Part 74:

OGAM Action Transmittal
U.S. DEPARTMENT OF HEALTH AND HUMAN SERVICES

Office of Grants and Acquisition Management (OGAM)
Office of the Assistant Secretary for Management and Budget
Room 517D – Hubert H. Humphrey Building
200 Independence Ave. S.W.
Washington, D.C. 20201

ACTION TRANSMITTAL – EXTERNAL

Transmittal No.: OGAM AT 2000–1
Date: November 15, 1999

TO: Federal Grantees and Awarding Agencies
SUBJECT: Changes in the Treatment of Research Costs Related to Animal Facilities

REGULATION: OMB Circulars A–21, A–122 and Appendix E, 45 CFR Part 74

APPLICABILITY: Federal Grantees and Awarding Agencies

EFFECTIVE DATE: Upon Issuance for All Newly Submitted Proposals for Facilities and Administrative Cost Rates

PURPOSE AND BACKGROUND: Office of Management and Budget Circulars and HHS regulations provide guidance on the treatment of specialized service facilities, including animal facilities, if material in amount. The animal care facilities of research institutions are required by OMB and Departmental regulations to be charged directly to Federal grants on a fee-for-service basis. This fee normally consists of both the direct costs and the allocable share of indirect costs (also known as Facilities and Administrative [F&A] costs) of the service. The purpose of this OGAM Action Transmittal is to clarify what facilities costs are to be considered part of the fee (and charged directly) and what portion should be treated and charged as an F&A cost. This clarification is required because, in recent years, the sophistication of animal research has caused more of this animal research to be conducted within the confines of these facilities. Since most nonanimal research takes place in office or laboratory space (which is included as part of the F&A cost), an inequity exists.

ACTION: Based on the changing nature of research conducted in these facilities, we are changing our methodology to include a certain portion of animal facility costs in the institution's F&A rates. This includes procedure rooms, operating and recovery rooms, isolation rooms, and quarantine rooms directly related to research protocols, as well as rooms that house animals involved in research that are not generally removed from the facility for conducting research. Notwithstanding this policy change, institutions must continue to document (through a space survey) the particular research projects conducted in research space included in an F&A pool.

In addition, to avoid potential over-allocations of F&A costs, on a case-by-case basis animal care charges may be treated like patient care costs and excluded from the allocation base used to charge F&A costs to awards.

To summarize, this Action Transmittal establishes a methodology for grantee organizations to account properly for costs of animal facilities.

AUTHORIZING OFFICIAL: Terrence J. Tychan Deputy Assistant Secretary for Grants and Acquisition Management

Summary of Findings from the Ohio State University – Committee on Institutional Cooperation Study (CIC)

The summary of findings published in this report may not reflect the opinions, policies, or practices of the individual institutions that participated in the study.

Cost-Recovery Approaches

1. Institutions recovered 20–76% of the total animal care costs through recharge mechanisms.

2. Participating institutions practiced different approaches to cost accounting for care of research animals.

3. Institutional funding of various components of animal care varied widely.

4. In most of the participating institutions, charges to investigators were only loosely related to underlying costs.

Operating Costs

1. Direct labor is the largest and most important factor in determining costs, representing 50–65% of the cost structure.

2. Labor performance improves with increasing program scale.

3. Labor performance tends to improve as activity is concentrated in fewer facilities or as facilities are used more intensively. As the average number of labor hours per animal housing room increases, the labor cost per animal decreases.

4. Labor performance tends to improve as activity is concentrated around fewer investigators or as average investigator activity increases.

5. Animal care programs with moderate scale and high complexity (many species and many services) have some structural explanations for higher costs.

6. Improving direct labor performance is a very effective way to reduce operating costs.

a) Reduce complexity by consolidating activity into fewer rooms and facilities wherever possible.

b) Focus on improving performance of animal care staff, through close measurement and management.

c) Reduce complexity of care (activities other than direct animal care) to help to reduce other costs for supplies and services, transportation, supervision, and protective clothing. Alternatively, the cost of complex services should be recovered outside the per diem charge.

Administrative and Indirect Costs

1. Complexity of animal care program administration can materially affect costs.

2. Animal purchasing and setup costs can have a substantial impact on short-term protocols and protocols that use expensive animals.

3. A mix of per diem and direct service charges makes good sense in that the user pays for special services. This mixture of charges assesses the true cost (assuming that the institution does not subsidize part of the animal care program from other institutional resources) of operations and maximizes the predictability of cost recovery. Accounting systems that roll these costs into their per diems are generally subsidizing short–term and complex projects or research with certain species at the expense of long–term and less complex projects.

Veterinary Staffing

1. Veterinary technicians, animal technicians, and veterinary residents can extend the capacity of the professional veterinary staff.

2. Number of investigators per veterinarian and number of protocols per veterinarian have little correlation across institutions.

3. Other surveys have found reasonable correlation between veterinary staffing and the number of nonrodent mammals in an institution. As the number of rodents grows, this correlation may decrease.

APPENDIX
C

Animal Resources Survey-1999 and Survey Tables

INTRODUCTION

This appendix contains the questionnaire that was sent to 130 animal care and use programs throughout the United States. The Committee on Cost of and Payment for Animal Research reviewed the questionnaire and suggested some enhancements that were incorporated into the survey by Yale Section of Comparative Medicine personnel before it was distributed. There were 63 responses for a nearly 50% response rate. The focus of the Cost Committee was to suggest methods for cost containment in traditional biomedical animal research facilities. Judging from the numbers and types of species used, some of the respondents to the survey appeared to be primarily in agricultural research or aquaculture. Therefore, the decision was made to restrict analysis to the 53 institutions that had an average daily mouse census of 1,000 or more. The 53 institutions were divided into three groups according to average daily mouse census: group 1 (n = 23) 1,000-9,999, group 2 (n = 16) 10,000-29,999, and group 3 (n = 14) > 29,999.

Group	Mouse average daily census	Institution ID numbers	No. institutions
1	1,000-9,999	4, 5, 6, 9, 12, 15, 17, 18, 20, 24, 28, 29, 34, 37, 39, 45, 46, 49, 53, 56, 57, 58, 59	23
2	10,000-29,999	11, 14, 19, 23, 25, 27, 36, 41, 42, 43, 44, 47, 54, 55, 60, 62	16
3	> 29,999	1, 3, 7, 10, 16, 21, 31, 35, 40, 48, 51, 52, 61, 63	14

The responses to the questionnaire are summarized in the ensuing tables. Nearly all tables have 1 row for each group and a final row for all 53 institutions. Where necessary, a description (in parentheses) of what the numbers in the table represent (mean number of institutions, mean percentage of the group or of all 53 institutions, and so on) is provided.

Animal Resources Survey – 1999

General Instructions

Please use black ink.

Please write legibly.

Please answer all questions.

Please do not add explanatory notes to your answers unless they are requested.

If you are unsure about the accuracy of a proposed answer (eg, institutional financial data), please ask an appropriate colleague at your institution for help.

If you are unsure about the intent of a question or how to answer a question, send your query by e–mail to: **valeria.krizsan@yale.edu.** We will try to help.

Please do not separate questionnaire pages. If you must do so, please restaple them securely before you return the questionnaire.

Please remember to enclose with the completed questionnaire your:

- organizational chart
- list of per diem rates
- financial contribution
- Please return the completed questionnaire by MARCH 15, 1999.

Questionnaire No. _____

Name of Institution_____

Private institution_____   Public institution_____

Name of unit for which data is being reported (eg, University, School of Medicine, etc)

_____

Name of animal resource

_____

Name and academic degrees of resource director

****

Person responsible for completing this form:

Name _____

Title (relevant to animal resource) _____

Telephone no. (   )_____   Fax no. (   )_____

E–mail address:_____

Mailing address:_____

_____

The data reported cover the fiscal year (*select one*):

_____ July 1, 1997 through June 30, 1998
_____ October 1, 1997 through September 1, 1998
_____ January 1, 1998 through December 31, 1998

1. Physical Plant:

A. Configuration:

Which configuration describes most accurately the layout of your resource:

1. Fully centralized: (all sites contiguous (under "one roof")) _____

2. Partially centralized: (one dominant site and one or more regional sites) _____

3. De–centralized: (multiple regional sites of approximately equal size) _____

4. Total number of sites _____

Is your institution pursuing centralization or consolidation
of animal resources to improve operating efficiency? (*Circle one*)   Y   N

B. Space allocation for full physical plant: No. Ft^2

    1. Animal rooms _____ _____

    2. Procedure rooms _____ _____

    3. Washing centers (including autoclaves, etc) _____ _____

    4. Food and bedding storage rooms _____ _____

    5. Laboratory animal medicine exam/treatment rooms _____ _____

    6. Operating rooms _____ _____

    7. Diagnostic laboratory rooms (path + micro + etc) _____ _____

    8. Administrative and faculty offices, library, etc _____ _____

    9. All other rooms _____ _____

    10. Corridors ****** _____

TOTALS _____ _____

<u>**Percent**</u> **of total space available for animal housing** _____
(Animal room ft² divided by total ft²)

C. Security:

C1. Physical Security:

Number of <u>sites</u> from **A4** protected by*:*

electronics (eg card reader) _____

keys _____

electronics and keys _____

C2. Environmental security:

Number of animal <u>rooms</u> from **B1** protected by:

automated environmental monitoring or controls _____

emergency power _____

D. Characteristics of individual sites:

*The size ranges in the following table are given in **gross square feet (gsf).** Your responses should indicate the total number of sites, rooms and machines per size range. Example: 3 sites at 5,000 gsf x 20 animal rooms/site = enter 3 under No. sites and 60 under No. animal rooms.*

Size of site (gsf) ➔	0–5,000	5,001–10,000	10,001–20,000	> 20,000	Total
No. sites					
No. animal rooms					
No. washing centers					
No. tunnel washers					
No. rack washers					
No. autoclaves					
No. procedure rooms					

F. Housing for MICE:

F1. Current housing conditions

Data in the following table represent conditions for the following period: Month_____Yr_____

Housing or husbandry condition	No. cages (avg daily census)	No. mice (avg daily census)
Conventional cages (no bonnets) with water bottles		
Conventional cages with autowater		
Microisolation cages with water bottles		
Microisolation cages with autowater		
Individually ventilated cages with water bottles		
Individually ventilated cages with autowater		
Total mouse cages		***************
Total mice	************	

Total ft^2 assigned to housing of mice_____

Mice/ft^2 of mouse housing space_____

E2. Recent or planned additions to housing for MICE

Status ⇨ ⇨	Completed since 1993	Under discussion	Designed	Under construction	Completion due (year)
Census capacity					
Gross ft^2					
Use of individually ventilated racks (1 = high, 2 = moderate, 3 = low, 4 = none)					
Washing center? (Y or N)					

F. Facilities for animal health services:

(If some rooms identified in the following table are multi–purpose (eg bacteriology and serology) please enter the combination of uses and relevant square footage in the space provided under "Combined use").

Function	No. of rooms	Total ft ^{2}
Examinations/ minor procedures		
Surgery (sterile)		
Post–operative recovery		
Diagnostic imaging		
Intensive care		
Pharmacy		
Necropsy		
Histotechnology		
Bacteriology/parasitology		
Serology		
Virology		
Clinical chemistry		
Combined use:		
(Should equal totals obtained by summating I.B.6–8) **Totals**		

Section II, beginning on the next page, focuses on staffing. In addition to your responses, please enclose an *organizational chart* that includes the institutional official(s) to whom the resource director reports.

II. Staffing

The position titles used in Section II may not correspond exactly to those used by your resource. Generic terminology has been used in this survey to help you make comparable choices .

A. Administrative staff:

Full–time equivalents is abbreviated in this and all subsequent queries as FTEs.
Example: If you have two assistant directors and each devotes 50% effort, enter 2 in the "number of persons" column and 1.0 in the "FTEs" column).

Position	Number of persons	FTEs	Degree(s) of current occupants			
			DVM	PhD	MBA	Other
1. Director						
2. Assoc/assist director						
3. Business manager						
4. Informatics specialist						
5. Purchasing agent						
6. Regulatory compliance officer						
Total managerial staff (1–6)			******************************** ********************************			
Total clerical staff			********************************			

B. Animal care staff:

B1. Composition of animal care staff

Position	Number of persons	FTEs	Number with AALAS certification – (specify levels)
1. Senior manager for animal care			
2. Assistant manager for animal care			

3. Regional supervisor for animal care			
4. Training coordinator			
Total manager/supervisor staff (1–4)			************************ ************************
5. Animal technologist			
6. Animal technician			.
7. Assistant animal technician			
Total technical staff (5–7)			************************

B2. Configuration of animal care staff

Enter the number which most closely indicates the configuration of your staff.

> **1** = all
> **2** = majority
> **3** = minority
> **4** = none

Internal (institutional employees)   _____

External (eg outsourced to a commercial firm)   _____

Unionized (technicians)   _____

Centralized (technicians report directly to senior supervisor/manager(s))   _____

Regional (regional staffs are led by supervisor who reports to a senior
            supervisor/manager).   _____

Other configuration _____

B3. Criteria for determining animal care staffing levels

> Quantified time–effort reporting   _____
> Qualitative assessments by animal care supervisors   _____
> Other_____   _____

B4. Wages and benefits for animal care staff

Standard work week (hours) _____

Starting hourly wage for an entry level technician (animal care/sanitation) _____

Current average annual salary for the animal technician staff _____

Current fringe benefit rate (in %) for an animal care technician's salary _____

Annual benefit days for a technician with 5 years of service:

Vacation days	_____
Sick days	_____
Paid holidays	_____
Other recess days	_____
Personal days	_____
Total annual benefit days	_____

B5. Recruitment of animal care staff

*Rank the following factors for their impact on **limiting** your resource's ability to recruit (**Table A**) and retain (**Table B**) new staff:*

(1 = high, 2 = moderate, 3 = low, 4 = none)

TABLE A

Recruitment factor	Manager/ Supervisor	Technician
Starting salary		
Earning potential		
Benefits		
Training and experience		
Job responsibilities		
Career opportunities		

Regional competition		
Location of resource		

TABLE B

Retention factor	Manager/Supervisor	Technician
Earning potential		
Benefits		
Career opportunities		
Regional competition		
Working conditions		

B6. Training of animal care staff *(Check all strategies in use)*

Training coordinator employed by animal resource   _____

Inhouse courses, including AALAS training   _____

Regional (multi–institutional) AALAS training   _____

Informal on–the–job training   _____

Computer–based training   _____

Participation in regional/national meetings   _____

Extended training on the production, biology
and use of genetically altered animals (*beyond that offered
in AALAS coursework*)   _____

Other_____   _____

B7. Productivity of animal care staff

Please indicate, in the table on the following page, your responses for staff productivity for ***mouse husbandry*** *in your* ***most efficiently configured housing site(s)****:*

For small mouse ("shoebox") cages	Change station used?(Y or N)	Interval (days) between cage changes	Average number of cages changed per technician per week
1. Conventional cage with water bottle			
2. Conventional cage with autowater			
3. Microisolation cage with water bottle			
4. Microisolation cage with autowater			
5. Individually ventilated cage with water bottle			
6. Individually ventilated cage with autowater			

C. Laboratory animal medicine staff:

C1. Composition of laboratory animal medicine staff

Example for completing the following table: *If 2 persons each devote half–time effort, enter 2.0 under "no. of persons" and 1.0 under "FTEs".*

Title	No. of persons	No. of FTEs	Degree(s) for each person	Specialty board(s)f or each person	No. of approved but unfilled positions
1. Clinician					
2. Pathologist					
3. Microbiologist					

Title	persons	FTEs	degrees	boards	unfilled
4. Virologist					
Total professional staff (1–4)			********	********	
6. clinical technologist					
7. Necropsy prosector					
8. Clinical pathology technologist					
9. Histotechnologist					
10. Microbiology technologist					
11. Virology/serology technologist					
Other					
Total technical staff			********	********	

C2. *Academic appointments for laboratory animal medicine professional staff*

Please indicate the number of members of your professional staff who hold academic appointments.

Rank	Director	Clinician(s)	Pathologist(s)	Other service faculty
Professor				
Assoc Professor				
Assist Professor				
Instructor				
Other rank				
None				

C3. Criteria for size and configuration of laboratory animal medicine staff

Judgment of the resource director and senior staff   _____

Review and approval by a faculty user group   _____

Review and approval by the institutional administration   _____

Budgetary priorities   _____

Other _____   _____

III. Animal Procurement and Census

Please enter data consistent with the reporting period checked on the identification page (Page 2). (Enter "U" for unknown)

Species	Average daily census	No. purchased/year	No. produced internally/ year	No. quarantine groups/year**
Mouse				
Rat				*************
Other rodent				*************
Rabbit				*************
Dog				*************
Cat	·			*************
Pig				*************
Sheep/goat				*************
Primate				
Amphibian				*************
Miscellaneous				*************
Totals				

** Quarantine should reflect animals procured from external <u>non–commercial</u> sources.

IV. Services

A. Services for mice:

A1. Husbandry for mice

Methods used to prevent or minimize exposure to infectious agents in mice.

Caging types used:	static microisolation cages	Y N
	individually ventilated cages	Y N
	cages with water bottles	Y N
	cages with autowater	Y N
	changed in a HEPA–filtered change station	%
Interval (days) between changes for static microisolation cages		Days
Interval (days) between changes for individually ventilated cages		Days
Type of bedding used for mice		
Treatment of bedding (1 = none, 2 = autoclaving, 3 = none)		1 2 3
Treatment of water (1= reverse osmosis, 2 = autoclaving, 3 = acidification, 4 = chlorination, 5 = none)		1 2 3 4 5
Treatment of feed (1 = none, 2 = autoclaving, 3 = pasteurization, 4 = irradiation, 5 = none)		1 2 3 4 5
Maximum number of mice permitted per small (shoebox) cage		
Number of cage racks in a typical mouse room		
What do you consider to be the minimum aisle width between racks?		Ft

A2. Cage sanitation

Item	Conventional cage	Microisolation cage	Ventilated cage
Washed in hot water only			
Washed in hot water and detergent			
Autoclaved after washing			

A3. Waste disposal

Source	Sanitary sewer	Sanitary landfill	Incinerator	Other
Soiled bedding				
Other nonhuman waste				
Carcasses				
Hazardous animal carcasses				

B. Animal technology services and revenue sources:

Please use the following key for entries: **R** = rodent (mouse or rat)
C = carnivore (dog or cat)
N = nonhuman primate

Item	Fully covered by per diem fees	Covered by per diem fees supplemented by institutional funds	Separate fee (not part of per diem fees)	Not available
Housing				
Husbandry				
Census taking				
Gnotobiotics				
Intramural transport of animals				
Cage sanitation and waste disposal				
Euthanasia				
Breeding colony management, including record–keeping				

item	covered	part–covered	separate	not available
Special supplies (gowns, gloves, etc)				
Animal identification (eg ear punching, tattooing)				
Weaning				
Rederivation (Cesarean or other)				
Blood and tissue collection, including tail biopsies				
Standardized therapeutic medication (eg treatment for pinworms)				
Administration of compounds/drugs during experimentation				
Restraint (chemical or physical)				
Feeding of special diets				
Other:_____				

C. Outsourcing of animals and/or services:

Please indicate institutional policies and practices for outsourcing animals and animal care services. Outsourcing is defined as animal housing, animal husbandry or animal health care provided by external sources (eg a private firm) either on campus or off–campus. Please enter the number corresponding to the percentage of average daily census for each species for which the corresponding outsourcing policy/practice is used.

> Key: **0** = none
> **1** = ≤ 25%
> **2** = 26–50%
> **3** = 51–75%
> **4** = > 75%

	Mice/rats	Rabbits	Dogs/cats	Nonhuman primates	Farm animals
Animal housing and care outsourced					
Only animal care outsourced					
Animal health care outsourced					
Outsourcing used primarily to save space					
Outsourcing used primarily to decrease operating costs					
Outsourcing used to protect animal health					
Outsourcing involves off–campus housing					
Outsourcing involves contracting of external personnel to provide on–campus services					

D. Laboratory animal medicine services:

(Enter one <u>or more</u> letters corresponding to the following species in the relevant box(es)):

R = rodent (mouse or rat)
C = carnivore (dog or cat)
N = nonhuman primate

Services	Fully covered by per diem fees	Covered by per diem fees supplemented by institutional funds	Separate fee (not covered by per diem fees)	Not available
Health assessment during quarantine				

Microbiological monitoring for infectious agents (serology, etc)				
Therapy for naturally occurring illness				
Therapy for iatrogenic illness				
Consultation about animal experimentation (planning grant proposals, anesthesia, etc)				
Anesthesia for experimentation (eg experimental surgery)				
Post–operative care				
Euthanasia				
Pathology for naturally occurring conditions				
Pathology for iatrogenic conditions				
Clinical chemistry for naturally occurring illness				
Clinical chemistry for iatrogenic illness				
Microbiological assessment of cell lines				

E. Research Services:

Please indicate all sources that apply. If your animal resource or comparative medicine program has a core lab for producing KO mice, check " animal resource program").

Service	Animal resource program	Other internal source	External vendor	Fully recharged to users	Partially/fully subsidized by institution
Production of polyclonal antibody					
Production of monoclonal antibody					

Targeted mutagenesis for mice (KO mice)					
Transgenesis for mice					
Cryopreservation of embryos or sperm					
Phenotyping of genetically altered animals					
Experimental surgery					
Other: (please list)					

F. Communications and administrative services:

Service	Operative	Planned	Not offered
Assistance in preparing grant applications using animals			
Interactive web site			
Animal ordering by users "on–line"			
E–mail user lists to disseminate information			
Newsletter			
Programmed meetings with user groups			
Comprehensive computer–based accounting system			

V. Prevalence of infectious agents

Please indicate, in the following table, the <u>current prevalence</u> of infectious agents in your MOUSE colonies. Prevalence should be given as the percent of mouse rooms in which the agent or serological evidence of the agent is present. If the percent is unknown, enter "U".

Infectious agent	Percent of <u>barrier</u> rooms	Percent of <u>non–barrier</u> rooms
Mouse adenovirus		
Mouse hepatitis virus		
Mouse parvovirus or MVM		
Mouse rotavirus		
Pneumonia virus of mice		
Sendai virus		
Theiler's MEV		
Mycoplasma species		
Helicobacter species		
Pinworms		

VI. Finances

A. Fees for ancillary animal care services:

A1. Procurement/setup fees.

Do you have animal procurement/setup fees ?   _____

The procurement fee is based on:
Percent of total $$ for animal order
Percent of total $$ for animal order up to a set maximum   _____
Percent of cost/animal up to a set maximum   _____
A standard charge per animal, per box or per order regardless
of the total amount of the order   _____

The setup fee is based on:
Fixed fee per cage   _____
Fixed fee per order   _____
Percent of the per diem rate for the species   _____

The following services are included in the procurement/ set up fees:
Placing animal orders _____
Verification of animal orders for regulatory compliance _____
Administrative check–in for new arrivals _____
Health check for new arrivals _____
Transportation to animal room _____
Uncrating and caging of new arrivals _____
Preparation of cage cards, census, other records _____

Do you have a cage purchase charge incremental to per diem fees?　　Y　N

This charge is based on:
Charge per cage _____
Percentage of a research project's animal budget _____

Do you have a shipping charge for preparing and shipping animals
from your institution to another site?　　　　　　Y　N

For rodent cages with low occupancy such as singly–housed mice:
The full per diem rate is charged
A reduced per diem rate is charged _____
If a reduced rate is charged, indicate the percent
　reduction compared to the full rate _____

B. Variations in per diem charges:

Indicate which conditions in the following table warrant a per diem rate or charge which differs from the standard rate for basic care.

Key:
R = mouse or rat
C = dog or cat
N = nonhuman primate

Condition	Increased per diem rate or supplemental charge	Reduced per diem rate
Large colonies (eg high volume users)		
Short–term housing		

Breeding females		
Barrier housing (eg autoclaved equipment and supplies, "sterile" technique for cage servicing)		
Housing and husbandry for hazardous infectious agents (BL2)		
Housing and husbandry for hazardous infectious agents (BL3)		
Housing and husbandry for hazardous chemical agents		
Quarantine of mice from non–commercial sources		
Quarantine of dogs or cats		
Quarantine of nonhuman primates		

Please enclose a copy of your institution's per diem rates for FY 98–99

C. Formulation of Per diem rates:

How often do you adjust per diem rates each year?	1X	2X	3X	4X
How often do you cost account each year?	1X	2X	3X	4X

Do you use cost accounting is used primarily as:
        a guide for rate setting?   Y   N
        the absolute determinant for rate setting?   Y   N

Do you use the *NIH Cost Analysis and Rate Setting Manual*   Y   N
for cost accounting and rate setting?

Based on your most recent cost accounting, indicate the contribution (%) of the following costs to your per diem rate for MICE:

Internal Indirect Cost Centers	%
Maintenance and repair	_____
General and administrative costs	_____
Transportation	_____
Cage washing and sanitation	_____
Laboratory services	_____
Health care	_____
Training	_____

Direct Cost Centers (continued)

Receipt/processing   _____

Technical services   _____

Husbandry   _____

Total   100.00

Do per diem rates for a given species subsidize the rate(s) for another species?   _____

Have any species been removed (or been targeted for removal) from your institution's research program because they are too costly to maintain?   _____

Please name the affected species   _____

D. Extramural funding:

Please indicate the total current extramural funding for biomedical research and training for the components of your institution? Provide figures for as many boxes as possible.

Type	Source	Funding for **all types** of biomedical research and training ($$millions)	Total funding for **animal–related** biomedical research and training ($$ million)
Direct	NIH		
	Other federal		
	All other		
	SUBTOTAL		
Indirect	NIH		
	Other federal		
	All other		
	SUBTOTAL		
	TOTAL		

E. Operating budget:

E1. Expense categories

Indicate which of the following <u>categories</u> of expenses are typically included in the DIRECT operating budget for your animal resource, irrespective of the source(s) of off–setting revenues.

> **1** = totally included
> **2** = partially included
> **3** = not included

Animal purchases (including purchase price, transportation, etc) _____
Salaries for directors, managers and supervisors _____
Salaries for veterinarians and other animal health professionals _____
Wages for technical staff (animal care, vet techs, dx lab techs, etc) _____
Animal care supplies (food, bedding, detergents, etc) _____
Personnel supplies (uniforms, shoes, gloves, etc) _____
Safety supplies and equipment _____
Rodent caging _____
Water bottles _____
Nonhuman primate caging _____
Transportation services (gas, oil, licenses, vehicle maintenance) _____
Informatics services and supplies (software, connect fees, etc) _____
Computer purchases _____
Capital equipment _____
Service contracts on fixed equipment _____
Service contracts on moveable equipment _____
Pharmaceuticals for animal health _____
Serological/microbiological monitoring _____
Staff training expenses _____
Travel (AALAS meetings, etc) _____
Facilities maintenance (painting, plumbing, electrical,etc) _____
Energy costs for heating and lighting animal rooms _____
Regulatory license and accreditation costs _____
IACUC costs _____

E2. Salary sources

Please indicate the current salary sources (as percent) for staff for each of the categories listed. If a staff position has more than one member, indicate the total percent under each column for all individuals in the position. (Example: If salaries for 2 of 4 clinical veterinarians are paid from per diem revenues, enter "50" in the "per diem revenues" column.

Staff position	Per diem revenues	Institutional funds	Fees for service	Research funds	Total FTEs
Director					
Associate/Assistant Director(s)					
Clinical veterinarian(s)					
Pathologist(s)					
Microbiologist					
Virologist					
Veterinary assistants/techs					
Diagnostic laboratory techn(s)					
Business manager					
Senior animal care manager(s)					
Animal care supervisors					
Animal care technicians					
Regulatory (compliance) personnel					

Have you requested or do you expect a change during the coming year in institutional support for any of the positions listed above? For example, do you expect institutional support for clinical veterinary salaries to increase or decrease in the coming year? If so, please indicate the change.

_____

_____

_____

E3. Deficit coverage

Institutional policy for handling year–end deficits in the animal resource operating budget includes:

Carried forward by the resource   _____

Covered by the institution   _____

Either or both mechanisms cited above may be used   _____

F. Institutional subsidy:

This section asks for information about the institutional subsidy for your animal resource. The definition of "subsidy" is likely to differ among institutions. *Please be as accurate as possible with your answers.* Options are provided to minimize potential uncertainty about the source or level of subsidy.

Please indicate the items that apply to the institutional subsidy for your resource.

Items	Yes	No	Uncertain
The resource receives an institutional subsidy			
The subsidy is negotiated annually			
The subsidy is applied only to specific pre–determined expenses			
The subsidy can be used as a discretionary account for the resource			
The subsidy offsets operating costs for specific species			
The subsidy is used, in part, to cover year–end operating deficits			
Operating costs to which the subsidy is typically applied are:	***	**	*********
Director's salary			
Salaries for other professional staff or faculty			
Purchase of fixed equipment			
Purchase of moveable equipment			
Purchase of supplies for animal care and/or health services			
Minor renovations (<$50,000)			
Major renovations (>$50,000)			

Facility maintenance (eg floors, walls, plumbing, electrical, etc)			
Diagnostic laboratory costs			
Program development (eg environmental enrichment, new management techniques, new diagnostic tests, informatics, etc)			
IACUC operations			
Veterinary costs associated with regulatory requirements			
Hazardous waste disposal			
AAALAC accreditation cost			
Occupational health and safety programs			

Please indicate the **subsidy for the fiscal year reported in the survey** for:

Direct operating budget:$ _____

Regulatory activities: _____

Renovations and equipment: _____

All other categories: _____

Total subsidy $ _____

**Total subsidy as % of direct operating expense
indicated in your responses to VI.E.1 (p. 24)** _____

G. Indirect cost recovery:

The current federally negotiated **indirect cost rate** for your institution is: _____%

The current federally negotiated cost rate for your animal resource, if it differs
from the institutional rate: _____%

The *status of implementation of OMB Circular A–21* at your institution is:

Full implementation since (give date) _____
Implementation is in progress since _____
Due to be completed by _____
Implementation is scheduled to begin by (give date) _____
There are no current plans for implementation _____

Institutional strategies for complying with A–21 include(d) which of the following:

        Increase fees to animal users   _____

         Designate animal resource space as organized research space   _____

        Subsidize the resource with institutional funds   _____

        The increased subsidy is/was :

                           Transient: _____

                         Expected to be permanent: _____

The ***estimated increase in per diem rates for MICE*** if the full cost is
absorbed by recharges is: _____%

The ***actual increase in per diem rates for MICE*** after institutional strategies
(indicated above) were activated was: _____%

The ***impact of A–21 implementation on animal census*** was:

                A permanent decrease in census   _____

                A transient decrease in census   _____

                Too early to tell   _____

VII. Regulatory Issues

Is your resource AAALAC–accredited?   _____

Approximately how many animal use protocols are active at any given time?   _____

Approximately how many ***full*** protocols are reviewed by the IACUC annually?   _____
(*Exclude annual updates and minor revisions*).

How many members serve on your IACUC?   _____

How many staff FTEs are employed by the IACUC?   _____

What is the estimated annual budget for the IACUC? $_____

Does your institution have a program for monitoring animal
experimentation apart from semi–annual IACUC inspections?   _____

If so, who conducts these inspections: _____

Please indicate the **compliance roles played by the staff/faculty veterinarians**.

Primary responsibility for:

        Initial review of every protocol   _____

        Initial review of selected protocols   _____

        Advising investigators on protocol preparation   _____

        training animal users   _____

How many FTEs are designated for meeting regulatory requirements
for training and monitoring of animal use?   FTEs

        Veterinarians   _____

        Other staff   _____

VIII. Resource–client Relationships

Please rank the following potential concerns among animal users at your institution:

1 = high level of user confidence and satisfaction
2 = most users are satisfied, but some are not
3 = general, moderate dissatisfaction
4 = substantial, widespread dissatisfaction and concern

Item	Rank
Per diem rates	
Animal procurement fees	
Space available for animal housing	
Quality and reliability of the physical plant	
Quality of animal care services	
Quality of laboratory animal medicine services	
Regulatory programs	
Training for animal users	
Institutional support for the resource	

The foregoing ranking is based on:

       Informal (anecdotal) information from users _____

       Formal survey of users _____

IX. Future Directions

Please **list up to 3 of the most important challenges facing your resource** in each of the following categories:

Physical Plant

Administration

Animal care services

Animal health services

Financial support

Regulatory compliance

Key to Survey Tables

Regulatory Program

Resource–client Relationships

Survey Responses

I. Physical Plant

I. A. Which configuration describes most accurately the layout of your resource:

Table 1. Physical plant: Configuration (number of institutions)

	Fully centralized	Partially centralized	Decentralized	Mean number sites/institutions	Centralization will increase
Group 1	3	18	2	6.13	6
Group 2	1	9	6	7.00	4
Group 3	1	10	3	11.71	3
All	5	37	11	7.87	13

I. B. Space allocation for full physical plant

Table 2a. Physical plant: Space allocation by number of rooms (mean number of rooms/institution)*

	Animal rooms	Procedure rooms	Washing centers	Food/bedding rooms	Exam/treatment rooms	Operating rooms	Dx lab rooms	Offices/library	Other rooms
Group 1	95.4	10.7	5.56	5.43	3.52	4.65	3.65	12.9	31.4
Group 2	113.6	19.2	10.81	9.56	3.25	8.00	4.75	19.8	39.4
Group 3	178.1	21.7	13.43	11.64	8.64	7.42	4.92	19.9	68.7
All	122.8	16.2	9.23	8.32	4.79	6.40	4.32	16.8	43.7

* Dx lab: Diagnostics laboratory

Table 2b. Physical plant: Space allocation by ft² (mean ft²/institution)*

	Animal rooms	Procedure rooms	Washing centers	Food/bedding rooms	Exam/treatment rooms	Operating rooms	Dx lab rooms	Offices/library	Other rooms	Corridors	Total ft²	% total ft² used as animal rooms
Group 1	24,931	1,935	3,144	1,395	701	1,463	938	2,149	10,226	9,489	56,757	44
Group 2	26,322	3,535	6,180	1,519	556	2,488	1,071	3,681	15,456	15,642	76,756	34
Group 3	38,052	3,157	6,648	2,938	1,388	2,163	1,503	3,700	12,032	11,553	84,468	45
All	28,639	2,733	4,954	1,819	828	1,953	1,120	3,008	12,287	11,898	70,114	41

* Dx lab: Diagnostics laboratory

Survey Tables – Page 1

Key:
Group (n)	#mice
1 (23)	1,000-9,999
2 (16)	10,000-29,999
3 (14)	>29,999

I. C. Physical security protection of all sites and environmental security protection of all animal rooms

Table 3. Physical plant: Security

	Mean number of sites/institution protected by:			Mean number of rooms/institution protected by:	
	Electronics	Keys	Electronics & keys	Environmental monitoring	Emergency power
Group 1	3.96	11.74	10.48	56.7	47.5
Group 2	2.38	2.19	3.06	87.6	88.7
Group 3	3.00	11.50	5.71	79.5	91.2
All	3.23	8.79	6.98	72.1	70.9

I. D. Characteristics of individual sites

Table 4. Physical plant: Characteristics of sites (mean number of sites or rooms/institution)

	Sites per size range					Total					
	0–5,000 ft²	5,001–10,000 ft²	10,001–20,000 ft²	> 20,000 ft²	Sites	Animal rooms	Procedure rooms	Washing centers	Tunnel washers	Rack washers	Autoclaves
Group 1	3.00	1.09	0.82	0.68	5.55	96.6	9.5	4.05	1.27	3.36	3.95
Group 2	3.56	1.62	1.44	0.62	7.25	114.1	17.2	7.25	1.88	5.50	7.56
Group 3	6.14	3.00	1.64	0.86	11.64	176.4	23.8	9.00	2.79	7.86	10.64
All	4.02	1.77	1.23	0.71	7.71	123.5	15.7	6.37	1.87	5.23	6.87

Survey Tables – Page 2

I. E. *Housing for mice*

Key:
Group (n)	#mice
1 (23)	1,000-9,999
2 (16)	10,000-29,999
3 (14)	>29,999

Table 5a. Physical plant: Current housing for mice (mean average daily cage census/institution)*

	Conv + bottles	Conv + autowater	MI + bottles	MI + autowater	IVC + bottles	IVC + autowater	Total cages	Total mice
Group 1	407	27	1,734	1.4	105	10	2,338	5,475
Group 2	910	312	4,235	56.2	733	616	6,868	21,705
Group 3	1,453	679	10,292	0	720	450	13,580	49,727
All	844	311	4,808	17.9	464	315	6,759	22,383

* Conv: conventional caging; MI: microisolette caging; IVC: individually ventilated cages

Table 5b. Physical plant: New or planned housing for mice (mean average daily cage census/institutions)*

	Completed since 1993				Under discussion				Designed			
	Gross ft²	Cage capacity	IVC use	Washing center	Gross ft²	Cage capacity	IVC use	Washing center	Gross ft²	Cage capacity	IVC use	Washing center
Group 1	10,467	7,764	3.00	6	6,616	5,504	2.83	5	50,400	3,250	2.50	2
Group 2	25,524	14,996	2.86	7	24,575	34,850	1.71	6	16,542	26,469	1.78	8
Group 3	15,032	10,516	3.31	10	16,479	11,750	2.22	7	46,070	19,396	2.67	4
All	16,272	47,229	3.11	23	14,442	17,665	2.23	18	30,899	25,356	2.17	14

* IVC: individually ventilated cages; 1 = High; 2 = Moderate; 3 = Low; 4 = None

Table 5c. Physical plant: Housing for mice under construction (mean average daily cage census/institutions)*

	Under construction			
	Gross ft²	Cage capacity	IVC use	Washing center
Group 1	18,333	20,933	1.00	2
Group 2	10,076	5,819	2.43	5
Group 3	27,220	16,282	2.33	5
All	16,981	12,330	2.12	12

IVC: individually ventilated cages; 1 = High; 2 = Moderate; 3 = Low; 4 = None

Survey Tables – Page 3

Key:
Group (n)	#mice
1 (23)	1,000-9,999
2 (16)	10,000-29,999
3 (14)	>29,999

I. F. Facilities for animal health service

Table 6a. Physical plant: Animal health facilities: Number of rooms (mean number of rooms/institution)

	Exams/ minor procedures	Surgery (sterile)	Post-op recovery	Diagnostic imaging	Intensive care	Pharmacy	Necropsy	Histo-technology	Microbiology	Serology	Virology	Clinical chemistry	Multiple use
Group 1	5.60	4.26	1.35	0.65	0.26	0.42	1.74	0.22	0.37	0.28	0.14	0.23	2.65
Group 2	8.62	6.62	1.11	1.06	0.33	0.67	1.83	0.43	0.46	0.39	0.27	0.32	1.04
Group 3	7.40	6.36	3.04	1.50	0.97	0.95	2.67	0.87	0.49	0.40	0.59	0.32	2.07
All	5.99	5.53	1.72	1.00	0.47	0.64	2.01	0.45	0.43	0.34	0.30	0.28	2.01

Table 6b. Physical plant: Animal health facilities: Square footage (mean ft²/institution)

	Exams/ minor procedures	Surgery (sterile)	Post-op recovery	Diagnostic imaging	Intensive care	Pharmacy	Necropsy	Histo-technology	Microbiology	Serology	Virology	Clinical chemistry	Multiple use	Total
Group 1	1,180	1,373	272	160	103	102	43	94	134	142	62	89	270	4,081
Group 2	1,748	2,249	183	157	38	102	370	133	126	108	54	67	238	5,519
Group 3	1,139	1,991	356	227	210	136	443	208	152	121	209	102	651	5,901
All	1,341	1,801	267	177	112	111	418	136	137	126	98	86	361	4,996

Survey Tables – Page 4

II. Staffing

II. A. Administrative Staff (FTE = Full time equivalent)

Table 7a. Staffing: Administrative staffing: Directorship (mean/institution)*

	Director					Associate/assistant director				
	No.	FTE(s)	DVM	DVM+	No DVM	No.	FTE(s)	DVM	DVM+	No DVM
Group 1	23	1.0	8	14	1	27	0.9	7	8	2
Group 2	15	0.9	6	6	3	15	0.8	7	2	3
Group 3	14	0.8	5	9	0	18	0.9	4	6	1
All	52	0.9	19	29	4	60	0.9	18	16	6

* DVM+: DVM plus PhD or MS or other masters' degree; No DVM: Degree other than DVM such as PhD or a masters' degree.

Table 7b. Staffing: Administrative staffing: Other administrative staff (mean/institution)

	Number of staff members						Number of staff FTEs					
	Business manager	Informatics specialist	Purchasing agent	Regulatory/ comp. Officer	Total managerial staff	Total clerical staff	Business manager	Informatics specialist	Purchasing agent	Regulatory comp. Officer	Total FTEs managerial staff	Total FTEs clerical staff
Group 1	0.86	0.36	0.70	0.26	4.04	1.93	0.70	0.20	0.53	0.21	2.93	1.56
Group 2	0.88	0.31	0.81	0.56	4.44	2.94	0.81	0.26	0.70	0.39	3.19	2.22
Group 3	0.71	0.71	1.04	0.71	5.36	4.14	0.68	0.23	0.99	0.64	4.14	3.64
All	0.83	0.44	0.82	0.47	4.51	2.82	0.72	0.22	0.70	0.38	3.33	2.31

Survey Tables – Page 5

Key:

Group (n)	#mice
1 (23)	1,000-9,999
2 (16)	10,000-29,999
3 (14)	>29,999

II. B. Animal care staff

II. B. 1. Composition of staff

Table 8a. Staffing: Animal care staff: Mean number of staff members/institution

	Senior manager	Asst. manager	Regional supervisor	Training coordinator	Total mgr./ spvsr. staff	Animal care technologist	Animal care technician	Asst. animal care technician	Total technical staff
Group 1	-.17	0.52	1.17	0.26	3.13	1.57	5.43	9.61	16.61
Group 2	-.00	1.19	2.50	0.56	5.31	1.75	10.00	11.12	22.88
Group 3	-.21	0.71	3.64	0.50	6.07	5.00	20.71	17.57	43.29
All	-.13	0.77	2.23	0.42	4.57	2.53	10.85	12.17	25.55

Table 8b. Staffing: Animal care staff: Mean staff FTEs/institution

	Senior manager	Asst. manager	Regional supervisor	Training coordinator	Total mgr./ spvsr. staff	Animal care technologist	Animal care technician	Asst. animal care technician	Total technical staff
Group 1	-.09	0.45	1.03	0.10	2.68	1.36	5.54	8.40	15.30
Group 2	0.96	1.19	2.03	0.41	4.58	1.69	9.84	9.41	20.93
Group 3	-.21	0.71	3.61	0.41	5.95	4.86	20.63	16.67	42.17
All	-.08	0.74	2.02	0.28	4.12	2.38	10.83	10.89	24.10

Table 8c. Staffing: Animal care staff: Percent of staff with AAALAS certification*

	Senior manager			Asst. manager			Regional supervisor			Training coordinator			Animal care technologist			Animal care technician			Asst. animal care technician		
	A	T	Tg	A	T	Tg	A	T	Tg	A	T	Tg	A	T	Tg	A	T	Tg	A	T	Tg
Group 1	13	9	61	4	30	0	9	39	35	0	4	13	35	26	39	52	2	0	2	13	39
Group 2	0	19	75	0	69	56	38	56	44	6	25	0	6	28	19	88	3	6	1	0	0
Group 3	0	7	86	21	0	36	11	1	1	0	43	0	46	1	64	2	3	7	71	21	0
All	6	11	72	8	34	26	18	64	58	4	26	29	46	29	46	40	3	4	1	11	17

* Percent of institutions in groups or total; percents do not total to 100 as some institutions had no certified staff in that category; A: ALAT; T: LAT; Tg: technologist.

Survey Tables – Page 6

Key:
Group (n)	#mice
1 (23)	1,000-9,999
2 (16)	10,000-29,999
3 (14)	>29,999

II. B. 2. Configuration of staff

Table 8d. Staffing: Animal care staff: Configuration of animal care staff* (number of institutions)

Score	Institutional				Outsourced				Unionized				Centralized†				Regional‡			
	1	2	3	4	1	2	3	4	1	2	3	4	1	2	3	4	1	2	3	4
Group 1	21	1	1	0	0	0	0	16	10	2	1	4	11	2	0	4	5	2	2	8
Group 2	15	1	0	0	0	0	0	12	9	3	3	3	7	1	2	3	6	2	1	3
Group 3	14	0	0	0	0	0	2	13	6	3	1	4	3	2	3	4	3	4	1	4
All	50	2	1	0	0	0	2	41	25	8	2	11	21	5	5	11	14	8	4	15

*Numbers do not sum to group or total as some responses were left blank. 1 = all; 2 = majority; 3 = minority; 4 = none
† techs report directly to senior manager; ‡techs report to regional supervisor who reports to center

II. B. 3. Criteria for determining animal care staffing levels

Table 8e. Staffing: Animal care staff: Criteria for staffing levels (number of institutions using that criterion)

	Time–effort reporting	Assessments by supervisors	Other
Group 1	12	17	6
Group 2	5	14	2
Group 3	8	12	4
All	25	43	12

II. B. 4. Wages and benefits for animal care staff

Table 8f. Staffing: Animal care staff: Wages and benefits for animal care staff

	Profile				Annual technician benefit days					
	Standard work week (hours)	Entry level hourly wage ($)	Average annual salary for tech ($)	Current fringe rate (%)	Vacation days	Sick days	Paid holidays	Recess days	Personal days	Total benefit days
Group 1	39.3	8.53	21,779	0.26	15.0	11.2	9.6	0.4	1.9	38.2
Group 2	39.7	9.35	22,096	0.28	16.5	12.8	9.4	0.5	1.2	40.3
Group 3	38.9	9.55	23,268	0.27	15.6	12.1	10.3	2.3	1.9	40.7
All	39.3	9.05	22,268	0.27	15.6	11.9	9.7	0.9	1.7	39.5

Survey Tables – Page 7

Key:

Group (n)	#mice
1 (23)	1,000–9,999
2 (16)	10,000–29,999
3 (14)	>29,999

II. B. 5. Recruitment and retention of animal care staff (managerial/supervisory) (Tables 8g–h)

Table 8g. Staffing: Animal care staff: Recruitment and retention of animal care staff (managerial/supervisory staff): Impact of recruitment factors (number of institutions)*

Rating	Starting salary				Earning potential				Benefits				Training & experience				Job responsibility				Career opportunity				Regional competition				Location of resource			
	1	2	3	4	1	2	3	4	1	2	3	4	1	2	3	4	1	2	3	4	1	2	3	4	1	2	3	4	1	2	3	4
Group 1	7	10	4	2	5	13	3	2	1	6	9	7	5	12	5	1	3	11	6	3	4	10	7	2	4	6	9	4	6	5	7	5
Group 2	2	10	3	1	6	5	4	1	1	2	3	10	7	4	4	1	0	5	8	3	0	5	10	1	6	4	5	1	2	5	7	2
Group 3	5	7	2	0	4	5	3	2	1	2	2	9	1	9	1	3	3	4	5	2	2	6	6	0	5	4	4	1	2	3	6	3
All	14	27	9	3	15	23	10	5	3	10	14	26	13	25	10	5	6	20	19	8	6	21	23	3	15	14	18	6	10	13	20	10

* 1 = high impact; 2 = moderate; 3 = low; 4 = no

Table 8h. Staffing: Animal care staff: Recruitment and retention of animal care staff (managerial/supervisory staff): Impact of retention factors (number of institutions)*

Rating	Earning potential				Benefits				Career opportunity				Regional competition				Working conditions			
	1	2	3	4	1	2	3	4	1	2	3	4	1	2	3	4	1	2	3	4
Group 1	3	13	5	2	2	4	10	7	3	12	7	1	4	7	7	5	2	7	7	7
Group 2	3	11	1	1	1	2	4	9	0	11	3	1	4	5	5	2	0	7	7	2
Group 3	4	6	4	0	1	3	1	9	4	4	5	1	3	3	8	0	2	2	9	1
All	10	30	10	3	4	9	15	25	7	27	15	3	11	15	20	7	4	16	23	10

* 1 = high impact; 2 = moderate; 3 = low; 4 = no

Survey Tables – Page 8

Key:	
Group (n)	#mice
1 (23)	1,000–9,999
2 (16)	10,000–29,999
3 (14)	>29,999

II. B. 5. Recruitment and retention of animal care staff (technical) (Tables 8i–j)

Table 8i. Staffing: Animal care staff: Recruitment and retention of animal care staff (technical staff): Impact of recruitment factors (number of institutions)*

Rating	Starting salary				Earning potential				Benefits				Training & experience				Job responsibility				Career opportunity				Regional competition				Location of resource			
	1	2	3	4	1	2	3	4	1	2	3	4	1	2	3	4	1	2	3	4	1	2	3	4	1	2	3	4	1	2	3	4
Group 1	10	6	6	1	7	8	7	1	2	6	10	5	6	11	4	2	2	11	9	1	3	13	5	2	5	6	8	4	6	4	8	5
Group 2	5	5	5	1	4	6	6	0	1	1	4	10	5	5	3	3	0	8	6	2	1	6	7	2	2	9	3	2	3	4	6	3
Group 3	4	6	3	1	5	6	2	1	0	2	3	9	3	5	2	4	3	4	5	2	2	7	4	1	4	4	5	1	2	3	5	4
All	19	17	14	3	16	20	15	2	3	9	17	24	14	21	9	9	5	23	20	5	6	26	16	5	11	19	16	7	11	11	19	12

* 1 = high impact; 2 = moderate; 3 = low; 4 = no

Table 8j. Staffing: Animal care staff: Recruitment and retention of animal care staff (technical staff): Impact of retention factors (number of institutions)*

Rating	Earning potential				Benefits				Career opportunity				Regional competition				Working conditions			
	1	2	3	4	1	2	3	4	1	2	3	4	1	2	3	4	1	2	3	4
Group 1	6	10	5	2	2	4	10	7	7	12	8	1	5	7	6	5	4	8	5	6
Group 2	3	7	6	0	1	2	5	8	2	7	5	1	2	9	3	2	0	8	7	1
Group 3	4	7	3	0	0	4	1	9	2	9	1	2	5	5	3	1	4	4	5	1
All	13	24	14	2	3	10	16	24	6	28	14	4	12	21	12	8	8	20	17	8

* 1 = high impact; 2 = moderate; 3 = low; 4 = no

Survey Tables – Page 9

II. B. 6. Training of animal care staff – check all strategies in use

Table 8k. Staffing: Animal care staff: Training of animal care staff (number of institutions)

	On-staff training coord-nator	Inhouse courses, including AALAS	Regional AALAS training	Informal on the job training	Computer based training	Regional/ national meetings	Extended training on production, biology, use of genetically altered animals	Other
Group 1	9	21	16	23	11	17	7	2
Group 2	10	15	10	16	9	16	4	2
Group 3	5	13	8	14	2	11	7	3
All	24	49	34	53	22	44	18	7

Survey Tables – Page 10

Key:

Group (n)	#mice
1 (23)	1,000-9,999
2 (16)	10,000-29,999
3 (14)	>29,999

II. B. 7. Productivity of animal care staff

Please indicate your responses for staff productivity for mouse husbandry in your most efficiently configured housing sites (for small mouse "shoebox" cages (Tables 8l-n)

Table 8l. Staffing: Animal care staff: Productivity of animal care staff (cage changes (technician x week)): Conv. cage*

	Conv. cage + water bottle			Conv. cage + autowater			
	Change station used		Change interval. (days)	Changes per tech per week	Change station used		Change interval. (days)	Changes per tech per week
	Yes	No			Yes	No		
Group 1	0	15	7.4	345.7	0	9	3.9	187.0
Group 2	1	6	5.7	814.0	0	3	4.5	671.0
Group 3	0	7	7.0	804.5	0	3	7.0	1,215.0
Avg 2 & 3	1	13	6.4	809.2	0	6	5.8	943.0
All	1	28	6.8	569.8	0	15	4.6	489.0

* Conv: conventional

Table 8m. Staffing: Animal care staff: Productivity of animal care staff (cage changes (technician x week)): MI cage*

	MI cage + water bottle			MI cage + autowater			
	Change station used		Change interval. (days)	Changes per tech per week	Change station used		Change interval. (days)	Changes per tech per week
	Yes	No			Yes	No		
Group 1	15	7	5.2	405.1	0	6	4.7	10.0
Group 2	14	2	4.6	930.0	3	0	5.8	700.0
Group 3	14	0	5.9	896.2	1	1	5.5	960.0
Avg 2 & 3	28	2	5.2	914.2	4	1	5.7	804.0
All	43	9	5.2	691.2	4	7	5.3	511.7

* MI: microisolette

Table 8n. Staffing: Animal care staff: Productivity of animal care staff (cage changes (technician x week)): IVC*

	IVC + water bottle			IVC + autowater			
	Change station used		Change interval (days)	Changes per tech per week	Change station used		Change interval (days)	Changes per tech per week
	Yes	No			Yes	No		
Group 1	9	2	6.6	234.2	3	1	7.0	200.0
Group 2	8	0	8.9	446.3	1	6	10.2	425.4
Group 3	9	0	8.9	366.0	1	2	10.5	503.5
Avg 2 & 3	17	0	8.9	403.8	2	8	10.3	448.8
All	26	2	8.1	343.6	5	9	10.0	416.8

*IVC: individually ventilated cage

Survey Tables – Page 11

Key:

Group (n)	#mice
1 (23)	1,000–9,999
2 (16)	10,000–29,999
3 (14)	>29,999

II. C. Laboratory animal medicine staff
II. C. 1. Composition of laboratory animal medicine staff (Tables 9a–b)

Table 9a. Staffing: Laboratory animal medicine staff (mean number of staff members/institution)

	Clinician	Pathologist	Microbiologist	Virologist	Clinical technologist	Necropsy prosector	Clinical pathology technologist	Histotechnologist	Microbiology technologist	Virology/ Serology technologist	Other
Group 1	2.3	1.0	0.6	0.1	1.0	0.1	0.3	0.2	0.2	0.1	0.5
Group 2	2.2	0.5	0.2	0.0	0.9	0.2	0.2	0.2	0.1	0.1	1.3
Group 3	3.5	1.3	0.2	0.0	1.8	0.4	0.2	0.7	0.5	0.5	0.9
All	2.6	0.9	0.4	0.0	1.2	0.2	0.2	0.3	0.3	0.2	0.8

Table 9b. Staffing: Laboratory animal medicine staff (mean number of staff FTEs/institution)

	Clinician	Pathologist	Microbiologist	Virologist	Clinical technologist	Necropsy prosector	Clinical pathology technologist	Histotechnologist	Microbiology technologist	Virology/ Serology technologist	Other
Group 1	1.7	0.4	0.0	0.0	0.7	0.1	0.2	0.2	0.1	0.1	0.5
Group 2	1.7	0.4	0.2	0.0	0.7	0.1	0.1	0.2	0.1	0.1	1.2
Group 3	2.1	0.8	0.0	0.0	1.5	0.3	0.2	0.5	0.4	0.4	0.9
All	1.8	0.5	0.1	0.0	0.9	0.2	0.2	0.2	0.2	0.2	0.8

Survey Tables – Page 12

Key:

Group (n)	#mice
1 (23)	1,000-9,999
2 (16)	10,000-29,999
3 (14)	>29,999

II. C. 2. Academic appointments of laboratory animal medicine professional staff

Table 9c. Staffing: Laboratory animal medicine staff: Academic appointments for laboratory animal medicine staff (%of institutions)

	Director						Clinician					
	Prof.	Assoc. prof.	Assist. prof.	Intsr.	Other rank	None	Prof.	Assoc. prof.	Assist. prof.	Intsr.	Other rank	None
Group 1	26	17	22	0	13	13	9	26	26	17	26	9
Group 2	12	31	12	0	12	12	0	25	31	12	12	25
Group 3	57	14	14	0	7	0	21	21	36	7	14	21
All	30	21	17	0	11	9	9	24	30	13	19	17

Table 9d. Staffing: Laboratory animal medicine staff: Academic appointments for laboratory animal medicine staff (%of institutions)

	Pathologist						Other service faculty					
	Prof.	Assoc. prof.	Assist. prof.	Intsr.	Other rank	None	Prof.	Assoc. prof.	Assist. prof.	Intsr.	Other rank	None
Group 1	17	13	22	4	4	0	0	9	9	0	0	9
Group 2	0	12	6	0	0	6	0	0	0	0	0	0
Group 3	7	29	7	14	0	0	0	0	0	0	7	0
All	15	12	19	4	8	2	0	4	4	0	2	4

Survey Tables – Page 13

Key:
Group (n)	#mice
1 (23)	1,000–9,999
2 (16)	10,000–29,999
3 (14)	>29,999

II. C. 3. Criteria for size and configuration of laboratory animal medicine staff (check all that apply).

Table 9e. Staffing: Laboratory animal medicine staff: Criteria for size and configuration of laboratory animal medicine staff (number of institutions)

	Judgment of resource director	Review & approval by faculty users	Review & approval by institution	Budgetary priorities	Other
Group 1	20	2	14	15	3
Group 2	14	2	15	9	0
Group 3	14	4	9	8	1
All	48	8	38	32	4

Survey Tables – Page 14

III. Animal Procurement and Census

Key:
Group (n)	#mice
1 (23)	1,000-9,999
2 (16)	10,000-29,999
3 (14)	>29,999

Please enter data with the reporting period checked on the identification page (Tables 10a–d)

Table 10a. Animal procurement and census: Animal census and annual procurement/internal production (number of animals)

	Mouse				Rat				Other Rodent			
	Avg. daily census	Purchased	Produced internally	Quarantine groups	Avg. daily census	Purchased	Produced internally	Quarantine groups	Avg. daily census	Purchased	Produced internally	Quarantine groups
Group 1	9,881.3	17,426.0	6,267.4	0	18,926.0	6,907.3	461.3	0	279.6	475.4	15.0	0
Group 2	19,855.6	34,722.8	24,042.4	0	1,510.0	12,734.4	454.1	0	201.6	692.4	104.0	0
Group 3	46,184.9	39,233.1	56,665.2	0	2,253.6	14,482.2	6,121.1	0	275.4	673.7	131.1	0
All	22,482.0	28,200.0	23,042.9	0	9,264.3	10,594.0	1,704.1	0	254.9	593.3	70.2	0

Table 10b. Animal procurement and census: Animal census and annual procurement/internal production (number of animals)

	Rabbit				Dog				Cat			
	Avg. daily census	Purchased	Produced internally	Quarantine groups*	Avg. daily census	Purchased	Produced internally	Quarantine groups	Avg. daily census	Purchased	Produced internally	Quarantine groups
Group 1	139.6	372.0	5.2	0	25.0	66.0	1.1	<1	34.7	45.7	2.7	1
Group 2	93.0	583.9	6.3	0	28.9	171.4	8.4	1	13.2	68.1	1.0	1
Group 3	181.2	926.1	2.1	0	62.9	151.0	1.7	0	14.1	24.7	4.7	0
All	136.5	582.4	4.7	0	36.2	120.2	3.5	<1	22.8	46.9	2.7	1

Table 10c. Animal procurement and census: Animal census and annual procurement/internal production (number of animals)

	Pig				Sheep/Goat				Nonhuman Primate			
	Avg. daily census	Purchased	Produced internally	Quarantine groups*	Avg. daily census	Purchased	Produced internally	Quarantine groups	Avg. daily census	Purchased	Produced internally	Quarantine groups
Group 1	15.1	191.0	1	1	14.0	27.4	4	0	64.4	29.3	0	2.1
Group 2	16.3	253.2	0	0	3.9	46.7	0	1	53.8	22.5	2.2	12.8
Group 3	12.3	233.5	0	0	18.9	67.1	0	0	73.6	42.0	1.4	3.4
All	14.7	220.9	<1	<1	12.2	43.7	2	<1	63.6	30.6	1.0	5.7

Key:
Group (n)	#mice
1 (23)	1,000-9,999
2 (16)	10,000-29,999
3 (14)	>29,999

Table 10d. Animal procurement and census: Animal census and annual procurement/internal production (number of animals)

	Amphibian				Miscellaneous				TOTALS			
	Avg. daily census	Purchased	Produced internally	Quarantin*e groups	Avg. daily census	Purchased	Produced internally	Quarantine groups	Avg. daily census	Purchased	Produced internally	Quarantine groups
Group 1	168.2	148.2	13.0	9.8	412.4	596.2	88.5	0	4,3407.8	26,284.6	5,744.4	20.7
Group 2	160.5	484.3	0	0	73.4	831.3	154.0	0	1,9745.9	47,104.6	24,684.1	63.3
Group 3	439.3	576.6	289.3	0	421.0	665.1	2,196.0	0	48,392.64	51,045.5	51,841.5	60.7
All	237.5	362.8	82.1	4.2	312.3	685.4	665.0	0	3,7581.3	39,110.5	23,638.6	44.1

* Quarantine should reflect animals procured from external non–commercial sources.

Survey Tables – Page 16

Key:

Group (n)	#mice
1 (23)	1,000–9,999
2 (16)	10,000–29,999
3 (14)	>29,999

IV. Services

IV. A. Services for mice

IV. A. 1. Husbandry for mice. Methods used to prevent or minimize exposure to infectious agents in mice (Tables 11a–b)

Table 11a. Services: Services for mice: Husbandry for mice (number of institutions)

	Caging used to prevent infection								% changed in change station	Cage change interval (days)	
	MI cage*		IVC*		Water bottles		Autowater			MI cage	IVC
	Yes	No	Yes	No	Yes	No	Yes	No			
Group 1	22	1	11	10	22	0	7	15	55	5.4	8.2
Group 2	16	0	12	4	16	0	10	5	76	4.6	8.9
Group 3	14	0	11	3	14	0	4	10	61	5.9	8.9
All	52	1	34	17	52	0	21	30	63	5.3	8.7

* MI: microisolette; IVC: individually ventilated cage

Survey Tables – Page 17

Key:
Group (n)	#mice
1 (23)	1,000-9,999
2 (16)	10,000-29,999
3 (14)	>29,999

Table 11b. Services: Services for mice: Husbandry for mice (continued)

	Bedding type	Number of institutions	Max. mice/cage	Typical no. cage racks/room	Minimum aisle width between racks (ft)
Group 1	Alphadri	1	5.0	6.0	3.0
	Aspen chips	3	5.0	3.7	2.8
	Beta chips	2	4.5	3.0	3.0
	Corn cob	5	4.6	2.8	3.2
	Hardwood chips	7	4.4	4.2	3.1
	Paper	2	4.0	3.5	3.0
	Sanichips	1	2.0	3.0	4.0
	Virgin paper	1	5.0	3.0	2.0
	Wood chips	1	6.0	2.0	4.0
Group 2	Alphadri	1	5.0	8.0	2.0
	Aspen chips	2	5.0	4.5	3.5
	Beta chips	2	4.5	6.5	1.5
	Carefresh	1	5.0	3.0	4.0
	Corn cob	2	4.5	3.0	2.8
	Ground corn cob	1	4.0	2.0	2.5
	Hardwood chips	2	4.5	3.5	3.5
	Shredded aspen	1	5.0	3.0	3.0
	Sanichips	1	2.5	3.8	2.8
	Wood chips	2	5.0	5.5	2.5
Group 3	Not specified	1	5.0	7.0	8.0
	Corn cob	8	5.0	3.9	3.2
	Hardwood chips	2	5.0	6.5	3.3
	Sanichips	3	4.0	4.7	3.3
All			4.7	4.1	3.1

Survey Tables – Page 18

Key:

Group (n)	#mice
1 (23)	1,000–9,999
2 (16)	10,000–29,999
3 (14)	>29,999

IV. A. 2. Cage sanitation

Table 11c. Services: Services for mice: Mouse cage sanitation (number of institutions)*

	Hot water only			Hot water and detergent			Autoclaving		
	Conv. cage	MI cage	IVC	Conv. cage	MI cage	IVC	Conv. cage	MI cage	IVC
Group 1	0	0	0	15	21	8	1	17	9
Group 2	2	2	2	11	14	10	2	15	10
Group 3	1	4	3	8	11	6	1	11	7
All	3	6	5	34	46	24	4	43	26

* MI: microisolette; IVC: individually ventilated cage

IV. A. 3. Waste disposal

Table 11d. Services: Services for mice: Waste disposal (number of institutions)

	Soiled bedding				Other non–human waste				Carcasses				Hazardous animal carcasses			
	Sewer	Landfill	Incinerator	Other	Sewer	Landfill	Incinerator	Other	Sewer	Landfill	Incinerator	Other	Sewer	Landfill	Incinerator	Other
Group 1	3	17	6	0	6	9	12	0	0	1	21	2	0	0	22	3
Group 2	5	12	5	2	4	12	4	2	0	1	15	3	0	0	12	3
Group 3	3	11	3	1	3	10	6	0	0	2	10	2	0	0	10	3
All	11	40	14	3	13	31	22	2	0	4	46	7	0	0	44	9

Survey Tables – Page 19

Key:

Group (n)	#mice
1 (23)	1,000-9,999
2 (16)	10,000-29,999
3 (14)	>29,999

IV. B. Animal technology services and revenue sources for rodents (Tables 12a–c), Carnivores (Tables 12d–f) and nonhuman primates (Tables 12g–l)

Table 12a. Services: Animal technology services and revenue sources: Rodents* (number of institutions)

Score	Housing				Husbandry				Census				Gnotobiotics				Internal transport				Cage sanitation				Euthanasia			
	1	2	3	4	1	2	3	4	1	2	3	4	1	2	3	4	1	2	3	4	1	2	3	4	1	2	3	4
Group 1	9	13	0	0	11	11	0	0	10	12	0	0	1	4	4	6	7	9	3	2	10	11	0	0	6	7	9	0
Group 2	10	5	1	0	10	6	0	0	10	6	0	0	5	2	1	4	4	4	8	0	9	6	1	0	4	4	8	0
Group 3	8	6	0	0	8	6	0	0	9	5	0	0	1	3	3	4	4	6	4	0	7	6	1	0	5	4	5	0
All	27	24	1	0	29	23	0	0	29	23	0	0	7	9	8	14	15	19	15	2	26	23	2	0	15	15	22	0

* 1 = per diem only; 2 = per diem + institution funds; 3 = separate fee; 4 = not available

Table 12b. Services: Animal technology services and revenue sources: Rodents* (number of institutions) (continued)

Score	Breeding				Special supplies				Animal ID				Weaning				Rederivation				Specimen collection			
	1	2	3	4	1	2	3	4	1	2	3	4	1	2	3	4	1	2	3	4	1	2	3	4
Group 1	1	7	11	3	7	10	6	0	8	4	8	1	3	4	12	1	0	0	13	4	1	2	20	0
Group 2	3	1	8	3	6	6	4	0	5	1	10	0	2	3	9	2	0	1	10	4	0	0	15	1
Group 3	1	1	9	2	6	5	3	0	4	2	7	1	2	1	10	1	1	0	12	1	2	0	12	0
All	5	9	28	8	19	21	13	0	17	7	25	2	7	8	31	4	1	1	35	9	3	2	47	1

* 1 = per diem only; 2 = per diem + institution funds; 3 = separate fee; 4 = not available

Table 12c. Services: Animal technology services and revenue sources: Rodents* (number of institutions) (continued)

Score	Routine medicine				Administer compounds				Restraint				Special diets				Other			
	1	2	3	4	1	2	3	4	1	2	3	4	1	2	3	4	1	2	3	4
Group 1	12	8	3	0	1	1	17	3	5	2	14	1	5	6	12	0	1	0	1	0
Group 2	4	6	6	0	0	1	15	0	1	1	12	0	2	5	8	1	1	0	2	0
Group 3	5	4	5	0	0	0	13	0	1	0	12	0	1	4	9	0	0	0	0	0
All	21	18	14	0	1	2	45	3	7	4	38	1	8	15	29	1	2	0	3	0

* 1 = per diem only; 2 = per diem + institution funds; 3 = separate fee; 4 = not available

Survey Tables – Page 20

Key:

Group (n)	#mice
1 (23)	1,000-9,999
2 (16)	10,000-29,999
3 (14)	>29,999

Table 12d. Services: Animal technology services and revenue sources: Carnivores* (number of institutions)

Score	Housing				Husbandry				Census				Gnotobiotics				Internal transport				Cage sanitation				Euthanasia			
	1	2	3	4	1	2	3	4	1	2	3	4	1	2	3	4	1	2	3	4	1	2	3	4	1	2	3	4
Group 1	8	12	0	0	10	10	0	0	9	11	0	0	1	4	1	0	5	6	8	4	10	10	0	0	4	5	11	0
Group 2	9	3	1	0	10	3	0	0	10	3	0	0	2	0	0	0	5	3	2	8	9	3	1	0	3	2	8	0
Group 3	7	5	0	0	6	6	0	0	8	4	0	0	1	2	0	0	4	4	4	4	7	5	0	0	5	3	4	0
All	24	20	1	0	26	19	0	0	27	18	0	0	4	6	1	0	14	13	14	16	26	18	1	0	12	10	23	0

* 1 = per diem only; 2 = per diem + institution funds; 3 = separate fee; 4 = not available

Table 12e. Services: Animal technology services and revenue sources: Carnivores* (number of institutions) (continued)

Score	Breeding				Special supplies				Animal ID				Weaning				Rederivation				Specimen collection			
	1	2	3	4	1	2	3	4	1	2	3	4	1	2	3	4	1	2	3	4	1	2	3	4
Group 1	1	3	3	5	6	9	4	0	9	6	5	0	4	2	3	3	0	0	5	5	1	3	14	0
Group 2	2	0	2	5	7	4	4	0	4	2	5	0	1	1	1	4	0	0	1	6	0	0	11	0
Group 3	1	1	4	2	4	4	0	0	2	4	4	1	1	3	2	1	0	0	2	4	1	1	6	1
All	4	4	9	12	17	17	8	0	15	12	14	1	6	6	6	8	0	0	8	15	2	4	31	1

* 1 = per diem only; 2 = per diem + institution funds; 3 = separate fee; 4 = not available

Table 12f. Services: Animal technology services and revenue sources: Carnivores* (number of institutions) (continued)

Score	Routine medicine				Administer compounds				Restraint				Special diets				Other			
	1	2	3	4	1	2	3	4	1	2	3	4	1	2	3	4	1	2	3	4
Group 1	9	7	3	0	1	1	14	3	5	2	11	1	6	7	7	0	0	0	0	0
Group 2	4	3	6	0	0	0	12	0	1	1	10	0	2	3	6	1	1	1	0	0
Group 3	4	3	4	0	0	0	11	0	1	0	10	0	0	3	8	0	0	1	0	0
All	17	13	13	0	1	2	37	3	7	3	31	1	8	13	21	1	1	1	0	0

* 1 = per diem only; 2 = per diem + institution funds; 3 = separate fee; 4 = not available

Key:

Group (n)	#mice
1 (23)	1,000–9,999
2 (16)	10,000–29,999
3 (14)	>29,999

Table 12g. Services: Animal technology services and revenue sources: Nonhuman primates* (number of institutions)

Score	Housing				Husbandry				Census				Gnotobiotics				Internal transport				Cage sanitation				Euthanasia			
	1	2	3	4	1	2	3	4	1	2	3	4	1	2	3	4	1	2	3	4	1	2	3	4	1	2	3	4
Group 1	3	6	2	0	6	4	0	0	5	5	0	0	0	4	0	2	5	4	1	0	6	5	0	1	2	2	7	1
Group 2	5	5	1	0	7	4	0	0	7	4	0	0	2	0	0	4	1	2	8	0	6	4	1	0	3	2	6	0
Group 3	5	6	0	0	5	6	0	0	6	5	0	0	1	3	0	3	2	4	5	0	5	5	1	0	4	3	4	0
All	13	17	3	0	18	14	0	0	18	14	0	0	3	7	0	9	8	10	14	0	17	14	2	1	9	7	17	1

* 1 = per diem only; 2 = per diem + institution funds; 3 = separate fee; 4 = not available

Table 12h. Services: Animal technology services and revenue sources: Nonhuman primates* (number of institutions) (continued)

Score	Breeding				Special supplies				Animal ID				Weaning				Rederivation				Specimen collection			
	1	2	3	4	1	2	3	4	1	2	3	4	1	2	3	4	1	2	3	4	1	2	3	4
Group 1	1	3	1	2	4	5	3	0	4	4	4	0	1	1	2	3	0	0	3	3	0	2	10	0
Group 2	1	0	2	4	6	4	1	0	4	1	5	0	1	0	2	4	0	0	2	5	0	0	9	0
Group 3	1	1	4	4	4	4	3	0	3	5	2	0	0	1	2	4	0	0	1	5	0	1	7	1
All	3	4	7	10	14	13	7	0	11	10	11	0	2	2	6	11	0	0	6	13	0	3	26	1

* 1 = per diem only; 2 = per diem + institution funds; 3 = separate fee; 4 = not available

Table 12i. Services: Animal technology services and revenue sources: Nonhuman primates* (number of institutions) (continued)

Score	Routine medicine				Administer compounds				Restraint				Special diets				Other			
	1	2	3	4	1	2	3	4	1	2	3	4	1	2	3	4	1	2	3	4
Group 1	5	4	2	0	0	0	9	2	3	1	6	1	4	4	4	0	0	0	0	1
Group 2	4	4	3	0	0	1	10	0	0	1	9	0	1	2	6	1	0	0	0	0
Group 3	3	4	2	0	0	0	10	0	0	1	9	0	0	3	6	0	0	0	0	0
All	12	12	7	0	0	1	29	2	3	3	24	1	5	9	16	1	0	0	0	1

* 1 = per diem only; 2 = per diem + institution funds; 3 = separate fee; 4 = not available

Survey Tables – Page 22

IV. C. Outsourcing of animals and/or services
Indicate institutional policies and practices for outsourcing. Enter the number corresponding to the percentage of average daily census for each species for which the outsourcing policy is used.

Data too sparse to summarize usefully.

IV. D. Laboratory animal medicine services for rodents (Tables 14a–c), carnivores (Tables 14d–f), and nonhuman primates (Tables 14g–i).

Table 14a. Services: Laboratory animal medicine services: Rodents* (number of institutions)

	Quarantine health assessment				Microbiological monitoring				Therapy: natural illness				Therapy: iatrogenic illness				Consultation			
Score	1	2	3	4	1	2	3	4	1	2	3	4	1	2	3	4	1	2	3	4
Group 1	4	9	8	0	10	8	4	1	11	8	4	0	4	6	13	0	8	12	2	0
Group 2	6	2	7	0	7	5	4	0	4	6	6	0	2	2	12	0	9	7	0	0
Group 3	5	4	5	0	7	5	2	0	6	4	4	0	1	2	11	0	6	8	0	0
All	15	15	20	0	24	18	10	1	21	18	14	0	7	10	36	0	23	27	2	0

* 1 = per diem only; 2 = per diem + institution funds; 3 = separate fee; 4 = not available

Table 14b. Services: Laboratory animal medicine services: Rodents* (number of institutions) (continued)

	Anesthesia				Post-op care				Euthanasia				Pathology: natural conditions			
Score	1	2	3	4	1	2	3	4	1	2	3	4	1	2	3	4
Group 1	1	1	20	1	2	2	17	1	8	4	10	0	10	7	5	0
Group 2	1	1	13	0	1	3	12	0	6	4	6	0	8	6	2	0
Group 3	0	1	13	0	1	0	11	1	5	3	6	0	4	8	2	0
All	2	3	46	1	4	5	40	2	19	11	22	0	22	21	9	0

* 1 = per diem only; 2 = per diem + institution funds; 3 = separate fee; 4 = not available

Table 14c. Services: Laboratory animal medicine services: Rodents* (number of institutions) (continued)

	Pathology: iatrogenic conditions				Clinical chemistry: natural illness				Clinical chemistry: iatrogenic illness				Microbiology on cell lines			
Score	1	2	3	4	1	2	3	4	1	2	3	4	1	2	3	4
Group 1	4	2	16	0	11	6	4	1	3	1	16	2	2	1	14	4
Group 2	1	3	12	0	5	4	7	0	0	0	14	0	1	0	10	3
Group 3	0	2	12	0	5	6	3	0	0	0	14	0	1	1	9	2
All	5	7	40	0	21	16	14	1	3	1	44	2	4	2	33	9

* 1 = per diem only; 2 = per diem + institution funds; 3 = separate fee; 4 = not available
Survey Tables – Page 23

Key:
Group (n)	#mice
1 (23)	1,000–9,999
2 (16)	10,000–29,999
3 (14)	>29,999

Table 14d. Services: Laboratory animal medicine services: Carnivores* (number of institutions)

Score	Quarantine health assessment				Microbiological monitoring				Therapy: natural illness				Therapy: iatrogenic illness				Consultation			
	1	2	3	4	1	2	3	4	1	2	3	4	1	2	3	4	1	2	3	4
Group 1	2	7	8	0	4	7	4	2	10	7	4	0	4	5	12	0	7	11	2	0
Group 2	4	2	6	0	3	1	5	0	4	3	6	0	2	1	10	0	9	4	0	0
Group 3	4	3	5	0	4	4	3	0	5	4	4	0	1	1	11	0	5	8	0	0
All	10	12	19	0	11	12	12	2	19	14	14	0	7	7	33	0	21	23	2	0

* 1 = per diem only; 2 = per diem + institution funds; 3 = separate fee; 4 = not available

Table 14e. Services: Laboratory animal medicine services: Carnivores* (number of institutions) (continued)

Score	Anesthesia				Post-op care				Euthanasia				Pathology: natural conditions			
	1	2	3	4	1	2	3	4	1	2	3	4	1	2	3	4
Group 1	1	2	18	0	2	3	15	0	5	3	12	0	8	6	6	0
Group 2	1	0	12	0	0	1	12	0	3	2	8	0	8	4	1	0
Group 3	0	1	12	0	1	0	12	0	5	1	7	0	3	7	3	0
All	2	3	42	0	3	4	39	0	13	6	27	0	19	17	10	0

* 1 = per diem only; 2 = per diem + institution funds; 3 = separate fee; 4 = not available

Table 14f. Services: Laboratory animal medicine services: Carnivores* (number of institutions) (continued)

Score	Pathology: iatrogenic conditions				Clinical chemistry: natural illness				Clinical chemistry: iatrogenic illness				Microbiology on cell lines			
	1	2	3	4	1	2	3	4	1	2	3	4	1	2	3	4
Group 1	4	2	14	0	9	6	5	0	3	2	14	1	2	1	7	6
Group 2	1	3	9	0	3	3	7	0	0	0	12	0	0	0	4	3
Group 3	0	1	12	0	4	5	4	0	0	0	13	0	1	1	4	3
All	5	6	35	0	16	14	16	0	3	2	39	1	3	2	15	12

* 1 = per diem only; 2 = per diem + institution funds; 3 = separate fee; 4 = not available

Survey Tables – Page 24

Key:	
Group (n)	#mice
1 (23)	1,000-9,999
2 (16)	10,000-29,999
3 (14)	>29.999

Table 14g. Services: Laboratory animal medicine services: Nonhuman primates* (number of institutions)

	Quarantine health assessment				Microbiological monitoring				Therapy: natural illness				Therapy: iatrogenic illness				Consultation			
Score	1	2	3	4	1	2	3	4	1	2	3	4	1	2	3	4	1	2	3	4
Group 1	2	5	7	0	3	6	3	1	5	5	5	0	2	3	10	0	4	8	2	0
Group 2	3	2	7	0	2	3	5	0	3	4	5	0	1	1	9	0	7	5	0	0
Group 3	3	3	6	0	4	4	2	0	4	4	4	0	1	1	10	0	4	8	0	0
All	8	10	20	0	9	13	10	1	12	13	14	0	4	5	29	0	15	21	2	0

* 1 = per diem only; 2 = per diem + institution funds; 3 = separate fee; 4 = not available

Table 14h. Services: Laboratory animal medicine services: Nonhuman primates* (number of institutions) (continued)

	Anesthesia				Post-op care				Euthanasia				Pathology: natural conditions			
Score	1	2	3	4	1	2	3	4	1	2	3	4	1	2	3	4
Group 1	0	1	15	0	1	1	13	0	2	2	10	0	4	5	5	0
Group 2	0	0	12	0	0	1	11	0	3	2	7	0	7	5	0	0
Group 3	0	1	11	0	0	0	12	0	4	1	7	0	2	7	3	0
All	0	2	38	0	1	2	36	0	9	5	24	0	13	17	8	0

* 1 = per diem only; 2 = per diem + institution funds; 3 = separate fee; 4 = not available

Table 14i. Services: Laboratory animal medicine services: Nonhuman primates* (number of institutions) (continued)

	Pathology: iatrogenic conditions				Clinical chemistry: natural illness				Clinical chemistry: iatrogenic illness				Microbiology on cell lines			
Score	1	2	3	4	1	2	3	4	1	2	3	4	1	2	3	4
Group 1	2	1	11	0	5	5	4	0	1	0	12	1	2	0	5	4
Group 2	0	3	9	0	4	4	4	0	0	0	11	0	0	0	3	4
Group 3	0	1	11	0	3	6	3	0	0	0	12	0	1	1	4	3
All	2	5	31	0	12	15	11	0	1	0	35	1	3	1	12	11

* 1 = per diem only; 2 = per diem + institution funds; 3 = separate fee; 4 = not available

IV. E. Research services
Please indicate all sources that apply (Tables 15a–l)

Table 15a. Services: Research services: Polyclonal antibody (% of institutions offering service)

	Animal resource program	Other internal source	External vendor	Fully recharged to users	Partially/fully subsidized
Group 1	30	30	39	48	13
Group 2	44	56	56	56	0
Group 3	57	29	50	71	0
All	42	38	47	57	6

Table 15b. Services: Research services: Monoclonal antibody (% of institutions offering service)

	Animal resource program	Other internal source	External vendor	Fully recharged to users	Partially/fully subsidized
Group 1	22	52	30	35	17
Group 2	31	56	50	50	0
Group 3	36	50	57	64	7
All	28	53	43	47	9

Table 15c. Services: Research services: Gene targeting for mice (% of institutions offering service)

	Animal resource program	Other internal source	External vendor	Fully recharged to users	Partially/fully subsidized
Group 1	0	43	39	26	22
Group 2	19	62	44	31	6
Group 3	14	64	7	36	29
All	9	55	32	30	19

Table 15d. Services: Research services: Transgenesis for mice (% of institutions offering service)

	Animal resource program	Other internal source	External vendor	Fully recharged to users	Partially/fully subsidized
Group 1	4	52	39	22	26
Group 2	25	69	50	25	12
Group 3	14	79	7	43	21
All	13	64	34	28	21

Survey Tables – Page 26

Key:

Group (n)	#mice
1 (23)	1,000-9,999
2 (16)	10,000-29,999
3 (14)	>29,999

Table 15e. Services: Research services: Cryopreserve mouse embryos or sperm (% of institutions offering service)

	Animal resource program	Other internal source	External vendor	Fully recharged to users	Partially/ fully subsidized
Group 1	9	22	30	9	22
Group 2	19	25	38	6	19
Group 3	14	64	21	43	7
All	13	34	30	17	17

Table 15f. Services: Research services: Phenotype genetically altered animals (% of institutions offering service)

	Animal resource program	Other internal source	External vendor	Fully recharged to users	Partially/ fully subsidized
Group 1	0	30	35	13	22
Group 2	25	62	31	25	25
Group 3	43	50	14	43	21
All	19	45	28	25	23

Table 15g. Services: Research services: Experimental surgery (% of institutions offering service)

	Animal resource program	Other internal source	External vendor	Fully recharged to users	Partially/ fully subsidized
Group 1	65	26	17	30	26
Group 2	56	56	25	38	19
Group 3	50	21	0	57	14
All	58	34	15	40	21

Table 15h. Services: Research services: Other (% of institutions offering service)

	Animal resource program	Other internal source	External vendor	Fully recharged to users	Partially/ fully subsidized
Group 1	9	0	0	9	13
Group 2	6	0	0	0	12
Group 3	7	0	0	7	7
All	8	0	0	6	11

Survey Tables – Page 27

IV. F. Communications and administrative services

Table 16. Services: Communications and administrative services (number of institutions)

	Grant application assistance		Interactive web site		On-line animal ordering		Email user lists		Newsletter		User group meetings		Computer-based accounting	
	Operative	Planned	Operative	Planned	Operative	Planned	Operative	Planned	Operative	Planned	Operative	Planned	Operative	Planned
Group 1	19	1	13	7	5	11	17	4	13	2	15	2	5	15
Group 2	15	0	6	9	3	6	8	4	8	2	11	1	5	11
Group 3	12	0	7	7	1	10	10	3	6	2	11	0	4	10
All	46	1	26	23	9	27	35	11	27	6	37	3	14	36

Survey Tables – Page 28

V. Prevalence of infectious agents

Please indicate the current prevalence of infectious agents in your mouse colonies

Table 17a. Services: Laboratory animal medicine services: Prevalence of infectious agents in mice (number of institutions with infection)*

Agent	Mouse adenovirus		Mouse hepatitis virus		Mouse parvovirus/MVM		Mouse rotavirus		Mouse pneumonia virus	
Type of room	Barrier	Non-barrier	Barrier	Non-barrier	Barrier	Non-barrier	Barrier	Non-barrier	Barrier	Non-barrier
All	1	2	3	22	2	15	1	6	0	6

* MVM: minute virus of mice

Table 17b. Services: Laboratory animal medicine services: Prevalence of infectious agents in mice (number of institutions with infection)*

Agent	Sendai virus		Theiler's MEV		Mycoplasma species		Helicobacter species		Pinworms	
Type of room	Barrier	Non-barrier	Barrier	Non-barrier	Barrier	Non-barrier	Barrier	Non-barrier	Barrier	Non-barrier
All	1	0	1	7	1	2	14	14	9	21

* MEV: murine encephalomyelitis virus

Survey Tables – Page 29

Key:
Group (n)	#mice
1 (23)	1,000-9,999
2 (16)	10,000-29,999
3 (14)	>29,999

VI. Finances

VI. A. Fees for ancillary care services

VI. A. 1. Procurement/setup fees. Do you have animal procurement/setup fees?

Table 18a. Finances: Fees for ancillary animal care services: Animal procurement/cage setup fees (number of institutions)

	Procurement fee based on:					Setup fee based on:		
	Procurement/set-up fees?	% total $ for animal order	% total $ for animal order to set up max	% cost/animal up to set max	Standard charge/animal/box/ order, regardless of total $	Fixed fee per cage	Fixed fee per order	% of per diem rate
Group 1	18	5	1	0	11	2	7	1
Group 2	10	1	1	0	8	1	1	1
Group 3	9	4	0	3	4	0	2	2
All	37	10	2	3	23	3	10	4

The following services are included in the procurement/setup fees.

Table 18b. Finances: Fees for ancillary animal care services: Animal procurement/cage setup fees (number of institutions) (continued)

	Services included in procurement/setup fees:						
	Placing animal orders	Verification for regulatory compliance	Admin check-in for new arrivals	Health check for new arrivals	Transportation to animal rooms	Uncrating, caging of new arrivals	Preparation: cage cards, census, other records
Group 1	16	15	14	13	14	14	15
Group 2	11	10	10	6	7	6	8
Group 3	10	10	10	9	9	9	9
All	37	35	34	28	30	29	32

Do you have a cage purchase charge incremental to per diem fees? If so, this charge is based on:
Do you have a shipping charge for preparing and shipping animals to another site?
For rodent cages with low occupancy such as singly-housed mice:

Table 18c. Finances: Fees for ancillary animal care services: Animal procurement/cage setup fees (Number of institutions) (continued)

		Purchase charge based on:			Rodent cages with low occupancy, such as singly-housed:		
	Cage purchase charge?	Charge per cage	% of animal budget for project	Shipping charge	Full per diem rate is charged	Reduced per diem rate is charged	% reduction
Group 1	3	2	0	13	14	3	48
Group 2	2	2	0	11	11	3	40
Group 3	3	1	1	10	7	4	46
All	8	5	1	34	32	10	45

Survey Tables – Page 30

VI. B. Variations in per diem charges:
Indicate which conditions warrant a per diem rate or charge which differs from the standard rate for basic care for rodents (Table 19a), carnivores (Table 19b), or nonhuman primates (Table 19c)

Key:
Group (n)	#mice
1 (23)	1,000–9,999
2 (16)	10,000–29,999
3 (14)	>29,999

Table 19a. Finances: Increases in per diem charges: Rodents (number of institutions)*

	Large colonies	Short-term housing	Breeding female mice	Barrier housing	Hazardous agents (BL2)	Hazardous agents (BL3)	Hazardous chemicals	Quarantine: mice
Group 1	0	0	6	14	11	8	9	3
Group 2	1	1	2	9	9	8	7	11
Group 3	0	1	2	10	8	8	7	8
All	1	2	10	33	28	24	23	22

* BL2: animal biosafety level 2; BL3: animal biosafety level 3

Table 19b. Finances: Increases in per diem charges: Carnivores (number of institutions)*

	Large colonies	Short-term housing	Breeding females	Barrier housing	Hazardous agents (BL2)	Hazardous agents (BL3)	Hazardous chemicals	Quarantine: dog/cat
Group 1	0	1	1	3	4	2	3	6
Group 2	0	0	0	0	2	2	2	4
Group 3	0	1	0	1	3	3	1	4
All	0	2	1	4	9	7	6	14

* BL2: animal biosafety level 2; BL3: animal biosafety level 3

Table 19c. Finances: Increases in per diem charges: Nonhuman primates (number of institutions)*

	Large colonies	Short-term housing	Breeding females	Barrier housing	Hazardous agents (BL2)	Hazardous agents (BL3)	Hazardous chemicals	Quarantine: NHP
Group 1	0	0	1	3	3	2	3	7
Group 2	0	0	0	0	1	1	2	5
Group 3	0	0	0	1	4	3	1	6
All	0	0	1	4	8	6	6	18

* BL2: animal biosafety level 2; BL3: animal biosafety level 3

Key:
Group (n)	#mice
1 (23)	1,000-9,999
2 (16)	10,000-29,999
3 (14)	>29,999

VI. C. Formulation of per diem rates
How often do you adjust per diem rates each year?
How often do you cost account each year?
Do you use cost accounting primarily as a guide for rate setting? The absolute determinant for rate setting? Do you use the NIH Cost Analysis and Rate Setting Manual for cost accounting and rate setting?

Table 20a. Finances: Formulation of per diem rates: Policies (number of institutions)

	Rate adjustments per year			Cost accountings per year				Cost acct. guides rates		Cost acct. determines rate		NIH Manual used		Cross subsidy between species		Any species targeted or removed because of high rates?		Affected species
	1	2	12	12	4	2	1	Yes	No	Yes	No	Yes	No	Yes	No	Yes	No	
Group 1	23	0	0	0	6	0	15	21	2	2	21	15	8	9	11	4	19	Nonhuman primates
Group 2	15	1	0	1	1	3	11	16	0	2	14	12	4	5	11	0	16	
Group 3	13	0	1	1	2	1	10	14	0	1	13	13	1	4	9	1	13	Sea turtles
All	51	1	1	2	9	4	36	51	2	5	48	40	13	18	31	5	48	

Based on your most recent cost accounting, indicate the contribution (%) of the following costs to your per diem rate for mice:

Table 20b. Finances: Formulation of per diem rates: Contribution of costs to per diem rate for mice (%)

	Maintenance & repair	General & administrative	Transportation	Cage washing & sanitation	Laboratory services	Health care	Training	Receipt/ processing	Technical services	Husbandry
Group 1	6	16	1	11	5	4	1	1	2	51
Group 2	8	17	1	16	3	6	1	1	1	47
Group 3	3	12	0	10	3	8	1	0	1	56
All	6	15	1	12	4	5	1	1	2	51

Key:	
Group (n)	#mice
1 (23)	1,000–9,999
2 (16)	10,000–29,999
3 (14)	>29,999

Please enclose a copy of your institution's per diem rates for FY98–99 (Tables 20c–d)

Table 20c. Finances: Formulation of per diem rates: Current per diem rates ($)

	Mouse		Mouse basic		Mouse full		Rat		Rat basic		Rat full	
	Per mouse	Per cage	Per mouse	Per cage	Per mouse	Per cage	Per rat	Per cage	Per rat	Per cage	Per rat	Per cage
Group 1	0.20	0.55	0.16	0.46	0.31	0.91	0.46	0.94	0.33	0.69	0.77	1.50
Group 2	023	0.53	0.55	0.54	1.48	0.88	0.38	0.98	0.77	0.80	0.93	1.27
Group 3	0.29	0.42		0.46		0.67	0.62	1.07		0.89	1.25	0.81
All	0.22	0.50	0.24	0.50	0.55	0.81	0.49	0.98	0.51	0.81	0.99	1.25

Table 20d. Finances: Formulation of per diem rates: Current per diem rates ($)(continued)

	Hamster		G Pig		Rabbit	Ferret	Cat	Dog	Primate	Primate small	Primate large	Sheep	Pig	Frog
	Per animal	Per cage	Per animal	Per cage										
Group 1	0.50	0.85	1.10	1.67	2.40	2.97	4.39	9.89	7.18	5.00	9.63	11.10	11.11	1.88
Group 2	0.38	0.98	0.96	1.44	1.86	2.58	4.93	7.30	6.19	3.55	8.69	11.02	9.79	0.89
Group 3	0.46	1.20	0.99	1.38	1.89	2.85	4.50	8.45	7.89	4.88	8.34	9.09	8.86	0.97
All	0.46	1.01	1.03	1.52	2.11	2.83	4.57	8.82	6.97	4.63	8.65	10.31	10.16	1.31

Key:
Group (n)	#mice
1 (23)	1,000-9,999
2 (16)	10,000-29,999
3 (14)	>29,999

VI. D. Extramural funding
Please indicate the total current extramural funding for biomedical research and training for the components of your institution.

Table 21a. Finances: Extramural funding: All types of research and training (in millions of dollars, mean)

	Direct				Indirect				
	NIH	Other federal	All other	Subtotal	NIH	Other federal	All other	Subtotal	Total
Group 1	39.4	17.8	28.7	82.1	11.2	1.9	2.9	18.6	100.3
Group 2	86.9	29.2	40.6	152.4	39.6	5.5	6.9	50.2	196.3
Group 3	97.2	23.3	46.8	150.2	48.2	8.3	15.6	69.7	213.6
All	70.5	23.3	37.0	123.7	30.9	4.8	7.5	42.4	160.8

Table 21b. Finances: Extramural funding: Animal–related research and training (in millions of dollars, mean)

	Direct				Indirect				
	NIH	Other federal	All other	Subtotal	NIH	Other federal	All other	Subtotal	Total
Group 1	12.6	3.9	5.1	20.5	4.9	0.7	0.5	7.6	33.7
Group 2	41.4	4.5	6.3	54.0	19.7	0.5	1.0	20.9	72.1
Group 3	48.6	4.9	9.4	60.2	22.4	1.8	1.8	25.6	81.2
All	33.1	4.4	6.7	43.4	14.9	0.9	1.0	17.2	59.9

Survey Tables – Page 34

Key:

Group (n)	#mice
1 (23)	1,000–9,999
2 (16)	10,000–29,999
3 (14)	>29,999

VI. E. Operating budget
VI. E. 1. Expense categories
Indicate which of the following categories of expense are typically included in the DIRECT operating budget for your animal resources, irrespective of the source(s) of off–setting revenues (Tables 22a–c)

Table 22a. Finances: Operating budget: Expense categories in DIRECT operating budget (number of institutions)*

Rating	Animal purchases			Salaries: director, managers/ supervisors			Salaries: veterinarians & related			Wages: technical staff			Animal care supplies			Personnel supplies			Safety supplies, equipment			Rodent caging		
	1	2	3	1	2	3	1	2	3	1	2	3	1	2	3	1	2	3	1	2	3	1	2	3
Group 1	16	2	5	9	11	3	10	10	3	17	4	2	21	1	1	21	1	1	19	3	1	20	2	1
Group 2	8	1	5	12	4	0	12	4	0	15	1	0	16	0	0	15	0	0	16	0	0	15	1	0
Group 3	10	2	2	6	8	0	8	6	0	13	1	0	14	0	1	14	0	0	14	0	.	13	1	0
All	34	5	12	27	23	3	30	20	3	45	6	2	51	1	1	50	1	1	49	3	1	48	4	1

* 1 = fully included; 2 = partially included; 3 = not included.

Table 22b. Finances: Operating budget: Expense categories in DIRECT operating budget (continued) (number of institutions)*

Rating	Water bottles			NHP caging			Transportation services			Informatics services/supplies			Computer purchases			Capital equipment			Fixed equipment contracts			Movable equipment contracts		
	1	2	3	1	2	3	1	2	3	1	2	3	1	2	3	1	2	3	1	2	3	1	2	3
Group 1	21	1	1	10	2	7	16	4	2	14	6	3	16	5	3	9	8	6	15	6	2	15	6	2
Group 2	16	0	0	9	2	1	14	0	1	15	1	0	14	2	0	7	3	5	15	0	1	16	0	0
Group 3	13	1	0	6	4	2	12	1	1	9	5	3	9	5	2	2	11	1	12	1	1	12	2	0
All	50	2	1	25	8	10	42	5	4	38	12	3	39	12	3	18	22	12	42	7	4	43	8	2

* 1 = fully included; 2 = partially included; 3 = not included.

Table 22c. Finances: Operating budget: Expense categories in DIRECT operating budget (continued) (number of institutions)*

Rating	Pharmaceuticals			Serological/ microbiological monitoring			Staff training			Travel			Facilities maintenance			Energy costs			Regulatory license accreditation			IACUC costs		
	1	2	3	1	2	3	1	2	3	1	2	3	1	2	3	1	2	3	1	2	3	1	2	3
Group 1	19	4	0	21	1	1	19	4	0	17	3	3	6	13	4	4	1	18	14	5	4	1	2	3
Group 2	14	0	1	14	2	0	14	2	0	14	2	0	9	4	0	1	1	13	13	2	1	3	4	15
Group 3	13	1	0	11	3	0	10	4	0	10	4	0	2	9	3	0	0	14	9	2	3	3	3	9
All	46	5	1	46	6	1	43	10	1	41	9	3	17	26	9	5	1	45	36	9	8	9	11	33

* 1 = fully included; 2 = partially included; 3 = not included.
Survey Tables – Page 35

Key:
Group (n)	#mice
1 (23)	1,000-9,999
2 (16)	10,000-29,999
3 (14)	>29,999

VI. E. 2. Salary sources

Please indicate the current salary sources (as percent) for staff for each of the categories listed. If a staff position has more than one member, indicate the total percent under each column for all individuals in the position (Tables 23a–g)

Table 23a. Finances: Operating budget: Salary sources (%)

	Director				Associate/assistant director			
	Per diem revenue	Inst. funds	Fees for service	Research funds	Per diem revenue	Inst. funds	Fees for service	Research funds
Group 1	18	78	2	2	24	69	0	6
Group 2	31	62	0	7	53	40	0	7
Group 3	36	54	0	9	48	44	3	5
All	27	67	1	5	40	53	1	6

Table 23b. Finances: Operating budget: Salary sources (%)(continued)

	Clinical veterinarian				Pathologist			
	Per diem revenue	Inst. funds	Fees for service	Research funds	Per diem revenue	Inst. funds	Fees for service	Research funds
Group 1	27	72	1	0	16	72	3	9
Group 2	56	39	3	2	25	54	0	21
Group 3	60	32	3	4	39	42	1	18
All	46	50	2	2	28	55	1	16

Table 23c. Finances: Operating budget: Salary sources (%)(continued)

	Microbiologist				Virologist			
	Per diem revenue	Inst. funds	Fees for service	Research funds	Per diem revenue	Inst. funds	Fees for service	Research funds
Group 1	0	100	0	0	0	100	0	0
Group 2	17	27	17	40	—	—	—	—
Group 3	68	3	2	27	8	64	12	16
All	32	36	7	25	8	64	12	16

Survey Tables – Page 36

Table 23d. Finances: Operating budget: Salary sources (%)(continued)

	Veterinary assistant/tech				Diagnostic laboratory tech			
	Per diem revenue	Inst. funds	Fees for service	Research funds	Per diem revenue	Inst. funds	Fees for service	Research funds
Group 1	67	31	2	0	42	26	8	23
Group 2	58	20	18	4	46	20	16	18
Group 3	66	10	24	0	60	24	11	4
All	63	21	14	1	51	24	12	13

Table 23e. Finances: Operating budget: Salary sources (%)(continued)

	Business manager				Senior animal care manager			
	Per diem revenue	Inst. funds	Fees for service	Research funds	Per diem revenue	Inst. funds	Fees for service	Research funds
Group 1	40	59	2	0	48	52	0	0
Group 2	61	39	0	0	69	25	0	5
Group 3	71	29	0	0	85	7	4	4
All	54	45	1	0	66	30	1	3

Table 23f. Finances: Operating budget: Salary sources (%)(continued)

	Animal care supervisor				Animal care techs			
	Per diem revenue	Inst. funds	Fees for service	Research funds	Per diem revenue	Inst. funds	Fees for service	Research funds
Group 1	55	43	0	2	76	24	0	1
Group 2	82	14	0	4	84	14	0	2
Group 3	88	6	1	5	90	7	1	2
All	74	22	0	3	82	16	0	1

Table 23g. Finances: Operating budget: Salary sources (%)(continued)

	Regulatory personnel			
	Per diem revenue	Inst. funds	Fees for service	Research funds
Group 1	25	69	3	3
Group 2	34	66	0	0
Group 3	31	69	0	0
All	30	68	1	1

Survey Tables – Page 37

VI. E. 3. Deficit coverage

Institutional policy for handling year-end deficits in the animal resource operating budget includes:

Table 24. Finances: Operating budget: Operating budget deficit (number of institutions)

	Carried forward	Covered by the institution	Either mechanism may be used
Group 1	8	11	3
Group 2	8	8	0
Group 3	6	5	3
All	22	24	6

VI. F. Institutional subsidy

Please indicate all that apply to the institutional subsidy for your resource.

Table 25. Finances: Institutional subsidy: Overview (number of institutions)*

Response	Received			Negotiated annually			Applied to targeted expenses			Used as discretionary account			Offsets costs for specific species			Cover year-end deficits		
	N	U	Y	N	U	Y	N	U	Y	N	U	Y	N	U	Y	N	U	Y
Group 1	0	1	21	12	1	9	11	1	10	16	2	4	20	1	1	10	3	8
Group 2	1	0	14	10	0	6	9	0	6	9	2	4	13	0	2	6	0	8
Group 3	1	0	13	6	1	7	4	1	8	12	0	2	13	1	0	7	1	4
All	2	1	48	28	2	22	24	2	24	37	4	10	46	1	3	23	4	20

* Y = Yes; N = No; U – Uncertain

Survey Tables – Page 38

Key:

Group (n)	#mice
1 (23)	1,000–9,999
2 (16)	10,000–29,999
3 (14)	>29,999

Operating costs to which the subsidy is typically applied are: (Tables 26a–b)

Table 26a. Finances: Institutional subsidy: Application to operating costs (number of institutions)*

Response	Director's salary			Professional staff/ faculty salaries			Fixed equipment			Movable equipment			Supplies			Renovations (<$50,000)			Renovations (>$50,000)			Facility maintenance			Diagnostic labs		
	N	U	Y	N	U	Y	N	U	Y	N	U	Y	N	U	Y	N	U	Y	N	U	Y	N	U	Y	N	U	Y
Group 1	4	1	17	2	1	18	12	2	8	12	1	9	12	1	9	8	2	12	11	1	10	11	2	9	12	2	8
Group 2	2	0	13	5	0	11	7	0	9	8	0	7	9	0	6	6	0	9	10	0	6	10	0	6	8	0	8
Group 3	4	0	8	4	0	10	8	0	6	7	0	7	8	0	6	6	0	6	8	0	6	4	0	10	9	0	5
All	10	1	38	11	1	39	27	2	23	27	1	23	29	1	21	22	2	27	29	1	22	25	2	25	29	2	21

* Y = Yes; N = No; U – Uncertain

Table 26b. Finances: Institutional subsidy: Application to operating costs (continued) (number of institutions)*

Response	Program development			IACUC operations			Regulatory services from veterinarians			Hazardous–waste disposal			AAALAC accreditation			Occupational health		
	N	U	Y	N	U	Y	N	U	Y	N	U	Y	N	U	Y	N	U	Y
Group 1	12	2	8	12	1	9	7	1	14	12	1	9	12	2	8	12	2	8
Group 2	9	1	6	8	0	8	5	0	11	11	0	5	8	0	8	9	1	6
Group 3	10	0	4	3	0	10	5	0	9	5	0	9	7	0	7	3	0	1
All	31	3	18	23	1	27	17	1	34	28	1	23	27	2	23	24	3	25

* Y = Yes; N = No; U – Uncertain

Please indicate the subsidy for the fiscal year reported in the survey for:

Table 27. Finances: Institutional subsidy: Subsidy for fiscal year reported (mean in thousands of dollars)

	For direct operating budget	For regulatory activities	For renovations & equipment	For all other categories	Total subsidy	Subsidy as % of direct operating expense
Group 1	471	20	48	23	616	45
Group 2	306	39	25	306	804	28
Group 3	318	59	121	51	841	20
All	381	36	61	116	727	33

Survey Tables – Page 39

VI. G. Indirect cost recovery

The current federally negotiated indirect cost rate for your institution and your animal resource (if different) is:

The status of implementation of OMB Circular A-21 at your institution:

Institutional strategies for complying with A-21 include:

Table 28a. Finances: Indirect cost recovery (%)*

| | Status of OMB circular A-21 implementation | | Institutional strategies for complying with A-21 | | |
	Federal indirect cost rate for institution	No current plans for implementation	Increase animal user fees	Designate animal resource space as organized research space	Subsidize resource with institutional funds
Group 1	50	22	43	22	48
Group 2	56	25	56	48	50
Group 3	57	21	43	43	43
All	54	23	47	32	47

* Only 1 institution in group 1 and 2 in group 3 had a different ICR for the animal resource

The estimated increase in per diem rates for mice if the full cost is absorbed by recharges:

The actual increase in per diem rates for mice after institutional strategies (indicated above) were activated was:

The impact of A-21 implementation on animal census was:

Table 28b. Finances: Indirect cost recovery (continued) (%)

| | Estimated increase in per diem rates for MICE if full cost is absorbed by recharges | Actual increase in per diem rates for MICE after compensatory institutional strategies were activated | Impact of A-21 implementation on animal census | | |
			Permanent census decrease	Transient census decrease	Too early to tell
Group 1	72	9	2	3	10
Group 2	59	20	0	2	6
Group 3	54	16	1	1	4
All	64	13	3	3	20

Survey Tables – Page 40

VII. Regulatory Program Issues

Table 29. Regulatory program: Overview *

	Resource AAALAC accredited (number of institutions)		Number of active animal use protocols	Number of full protocols reviewed annually by IACUC	Number of members serving on IACUC	Staff FTEs employed by IACUC	Annual budget for IACUC ($)	Program for monitoring animal experimentation apart from semi-annual IACUC inspections?	
	No	Yes						No	Yes
Group 1	2	21	660	206	14	1.1	62,728	7	16
Group 2	1	15	425	400	16	1.9	85,928	4	12
Group 3	0	14	608	380	21	2.5	164,295	1	12
All	3	50	575	310	16	1.8	97,810	12	40

* AAALAC: Association for Assessment and Accreditation of Laboratory Animal Care; IACUC: institutional animal care and use committee

Please indicate the compliance roles played by the staff/faculty veterinarians.

Primary responsibility for:

How many FTEs are designated for meeting regulatory requirements for training and monitoring of animal use?

Table 30. Regulatory program: Staff duties and responsibilities

	Initial review of every protocol	Initial review of selected protocols	Advise investigators on protocol preparation	Train animal users	FTEs for training & monitoring animal use	
					Veterinarians	Other staff
Group 1	18	5	23	22	0.9	1.0
Group 2	14	5	15	13	3.6	1.0
Group 3	12	6	12	13	2.1	0.9
All	44	16	50	48	1.9	0.9

Key:
Group (n)	#mice
1 (23)	1,000-9,999
2 (16)	10,000-29,999
3 (14)	>29,999

Survey Tables – Page 41

Key:

Group (n)	#mice
1 (23)	1,000–9,999
2 (16)	10,000–29,999
3 (14)	>29,999

VIII. Resource–client Relationships

Please rank the following potential concerns among animal users at your institution.

Table 31. Resource–client relationships (number of institutions)*

	Animal user concerns																																				Ranking based on	
	Per diem rates				Animal procurement fees				Animal housing space				Quality/reliability of physical plant				Quality of animal care services				Quality of lab animal medicine services				Regulatory programs				Training for animal users				Institutional support for resource				Informal survey	Formal survey
Rating	1	2	3	4	1	2	3	4	1	2	3	4	1	2	3	4	1	2	3	4	1	2	3	4	1	2	3	4	1	2	3	4	1	2	3	4		
Group 1	1	12	7	3	0	9	13	1	2	6	11	4	2	7	10	4	1	12	10	0	0	17	6	0	5	14	4	0	0	9	13	1	0	5	12	6	19	5
Group 2	1	7	4	4	4	3	7	2	4	6	4	2	0	7	7	2	0	9	7	0	0	13	2	1	2	9	4	1	1	7	7	1	0	3	8	2	14	3
Group 3	2	10	0	2	2	5	5	2	1	2	8	3	0	4	10	0	0	6	6	2	0	9	4	1	2	8	4	0	0	5	8	1	0	3	8	1	12	3
All	4	29	11	9	6	17	25	4	7	14	23	9	2	18	27	6	1	27	23	2	0	39	12	2	9	31	12	1	1	21	28	4	0	11	28	9	45	11

* 1 = high; 2 = moderate; 3 = fair; 4 = poor

Biographical Sketches of
Committee Members

Christian E. Newcomer, Chair. Dr. Newcomer is Director of the Division of Laboratory Animal Medicine and Research Associate Professor of the Department of Pathology and Laboratory Medicine of the University of North Carolina. Dr. Newcomer is the immediate past president of the American College of Laboratory Medicine and Vice President of the Council on Accreditation, Association for Assessment and Accreditation of Laboratory Animal Care (AAALAC) International. His research interest is the infectious diseases of laboratory animals.

Frederick W. Alt is a Howard Hughes Medical Institute Investigator, Charles A. Janeway Professor of Pediatrics and Professor of Genetics at Harvard Medical School and Children's Hospital, and a Senior Investigator at the Center for Blood Research in Boston. He studies the molecular and cell biology of immunity. He sits on the editorial boards of Molecular and Cellular Biology, International Immunology, Developmental Immunology, Advances in Immunology, Current Biology, Science, and Immunity. He is a Co-Editor of *Current Opinion in Immunology*, an Advisory Editor for *Journal of Experimental Medicine* and a Contributing Editor for *Molecular Medicine*. He is a member of the National Academy of Sciences, the American Academy of Microbiology, and the American Academy of Arts and Sciences.

Ransom L. Baldwin is Professor and Sesnon Chair of the Department of Animal Science of the University of California at Davis. His research interests are in ruminant digestion, physiology of lactation, nutritional energetics, mechanisms and quantitative aspects of regulation of animal and tissue metabolism, and computer simulation modeling of animal systems. He was a member of the ILAR Guide Committee.

John Donovan is Vice President of Laboratory Animal Science and Welfare, Aventis Pharmaceuticals, Inc. From 1986 to 1994, he was Director of the Office of Laboratory Animal Science at the National Cancer Institute, National Institutes of Health. He is a Diplomate of the American College of Laboratory Animal Medicine (ACLAM) and was President of ACLAM 1994-5.

Janet Greger is Professor of Nutritional Sciences and Environmental Toxicology of the University of Wisconsin. She was both Associate Dean for Research of the Medical School and Professor of Nutritional Sciences and Environmental Toxicology of the University of Wisconsin, has chaired the all campus animal care and use committee at the University of Wisconsin and is on the Board of Trustees of AAALAC (1992-2000), serving on their strategic planning committee in 1996. She was also on the Board of Directors of the Council on Government Relations and was a member of the NRC committee that wrote the report on *Nutrient Requirements of Laboratory Animals*, fourth edition.

Joseph Hezir is a Managing Partner of the EOP Group, Inc., and was a cofounder of the Group. He was associated with Office of Management and Budget for 18 years, ending there as Deputy Associate Director for Energy and Science. He specializes in regulatory strategy development and problem solving, and identifying newly created government business opportunities formed from mergers, acquisitions, joint ventures, and new markets.

Charles McPherson is Executive Director of the American College of Laboratory Animal Medicine and an independent consultant in laboratory animal medicine. He was Chair of the Committee on Revision of Cost and Rate Setting Manual for Animal Research Facilities. He has been a leader in laboratory animal medicine and has published extensively on the care and use of laboratory animals.

Josh Steven Meyer is the managing principal of GPR Planners Collaborative, Inc., and a Registered Architect in the State of New York. Mr. Meyer has participated in the programming and planning of 60 major research projects and more than 40 animal facilities for academic, institutional and corporate clients. His assignments include existing facilities analysis, facilities master planning, and macro- and micro-level development of laboratory, pilot plant, and animal and toxicology facilities.

Robert B. Price is Executive Vice President for Administration and Business Affairs of the University of Texas Health Center. He has an extensive background in higher education, having held various positions at Texas Tech University, The University of Texas at Arlington, and the Health Science Center at San Antonio. He also was a member of the Board of Directors of the Council on Government Relations 1979-1986 and is currently Chairman of the Board.

Daniel H. Ringler is Professor and Director of the Unit for Laboratory Animal Medicine, University of Michigan Medical School. His research interests are: spontaneous diseases of laboratory animals, comparative medicine and management of research animal resources. He has served on and chaired the Council on Accreditation of the Association for Assessment and Accreditation of Laboratory Animal Care International. He has also served as president of the American College of Laboratory Animal Medicine and was a member of the Council of the Institute for Laboratory Animal Research.

James R. Swearengen is Director of the Veterinary Medicine Division of the U.S. Army Medical Research Institute of Infectious Diseases. He has extensive experience in directing multi-species animal care and use programs, supporting medical and surgical research and interfacing with scientific investigators. He has been involved in designing and providing oversight for the construction of animal care and research facilities.

John Vandenbergh is a Professor, Department of Zoology, North Carolina State University. His research areas are environmental control of reproduction, the endocrine basis of behavior, and rodent and primate behavior. He was a member of the committee to revise the *Guide for the Care and Use of Laboratory Animals* and has been on review panels for NSF and NIH. He is a member of the American Society of Zoologists, Animal Behavior Society (President 1982-83), and Society for the Study of Reproduction.